. . . For the safety of occupants, it has become clear that autos need a passive protection system.

"Passive" means that the system must require no effort by the passengers, and "protection" means that they must be saf   -
sion with a stationary barrier at 30 miles

The electro-mechanical system consists of si   g a sensor. Slightly larger than a golf ball   firewall. Its function is to determine when   ing and a metal weight are located next to the sensor and combine with it to form an electrical circuit. Mounted behind the instrument panel is a bottle of nitrogen gas which has a high-explosive cap. A tube connects the bottle to a coated-nylon bag folded up in the instrument panel. . . .

As diagrammed in Figure 1, the air bag activates when an impact forces the sensor against the spring. If the impact equals a barrier crash of 8 miles per hour, the spring moves the metal weight approximately 1/2 inch, completing the electrical circuit. Electric wires carry this signal to the nitrogen bottle's high-explosive cap. When the cap explodes, the nitrogen rushes out into the pipe and is quickly distributed to the air bag. . . .

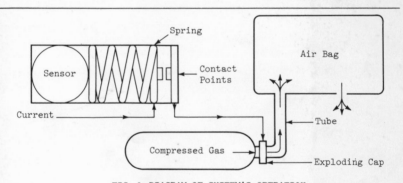

FIG. 1 DIAGRAM OF SYSTEM'S OPERATION

According to Robert Lund, the system activates, inflates the air bag, and deflates the bag in less than a second from the time of impact.[1] . . .

The need for a passive protection system in automobiles is shown in Table 1. Despite the fact that safety belts are now standard equipment, over 50,000 Americans are killed each year in auto accidents, and another 2 million are injured. These numbers could be reduced if safety equipment were worn, but the statistics show that few people wear it. Air bags could save at least 40 percent of the lives now lost in frontal collisions. . . .

TABLE 1   DATA ON AUTO OCCUPANTS

Occupants using shoulder harnesses	5%
Occupants using seat belts	30%
Injuries per year	2 million
Deaths per year	54,000

TECHNICAL REPORT WRITING TODAY

STEVEN E. PAULEY

Purdue University

HOUGHTON MIFFLIN COMPANY · BOSTON
Atlanta · Dallas · Geneva, Illinois · Hopewell, New Jersey
Palo Alto

For my father, Orrin F. Pauley

Printed in the U.S.A.

Library of Congress Catalog Card Number: 72-7923
ISBN: 0-395-12664-9

CONTENTS

Models

Proposal for Protection Against Spread of Alpha Radioactivity / 171

Proposal for Installation of a Power Generator / 180

Writing Assignments / 190

PREFACE

A technical writing course brings together specialists from many disciplines. The various technologies are represented by students specializing in career fields, and the course is guided by a writing specialist. The objective of the course is to improve the students' written communication, and that objective cannot be met unless communication exists between the technical specialists and the writing specialist. *Technical Report Writing Today* is designed to achieve that communication.

To allow the students and the instructor to meet each other halfway, *Technical Report Writing Today* advocates that students write papers about their technical areas and aim their writing at an uninformed reader. With this system, students have the advantage of selecting topics they are already familiar with, but they assume the responsibility of communicating their technical information to the instructor, who will understand the papers only if they present information clearly and contain explanations of technical terminology.

Writing for an uninformed reader is not an entirely academic exercise. Nobody in industry asks for reports telling him what he already knows, or complains that reports explain things too clearly, in language too simple. Students generally learn about their technical areas while writing papers of this type, and the instructor is able to make incisive, meaningful comments about the students' writing problems.

The model reports in this text, all but two of which were written by freshmen and sophomores, demonstrate that the writing assignments are applicable to all technical areas. The reports are imperfect, but their organization generally makes up for their weaknesses and gives students a pattern to follow. Students should examine the reports critically; in fact, they should be encouraged to debate the reports' strengths and weaknesses. The instructor's role is to insure that these discussions produce specific suggestions for improving the models, and that the students apply those suggestions to their own reports.

Many of the model reports contain visuals, but students should not expect visuals to do their communicating for them. The function of a visual is to reinforce words, but when it is substituted for good writing, it simply wastes space. Also, students who have not had a drafting course should not be concerned; a two-dimensional sketch can communicate more information than an ineffectively used three-dimensional drawing.

This text lends itself to both basic and advanced writing courses. Freshmen and sophomores generally devote most of the term to becoming skilled in the basic techniques of technical writing. For a final paper, they apply the techniques to a formal report, a research paper, or to an assignment that integrates the two. Juniors and seniors generally concentrate on writing letters and formal reports. In the type of course advocated by this text, advanced students are assigned problem-oriented reports on topics from their technical areas. They create hypothetical problems and attempt to solve the problems in reports aimed at an uninformed reader.

I wish to acknowledge the following students who have given me permission to print their reports in this text:

Jerry Bisch: "Stress and Strain"

Michael Bartkowicz: "Dynamics"

Earl H. Phillips: "Audio-Visual Synaesthesia"

James R. Schroeder: "Chemical Equilibrium"

James W. Jahn: "Description of an Automatic Center Punch"

James Pause: "Description of a One-Inch Micrometer"

Eric D. Cronch: "Description of a Moving-Coil Meter"

Daniel W. Stolarz: "Description of a Long-Nosed Pliers"

John W. Pio: "Description of an Ammeter in Operation"

John A. Predaina: "Operation of the Basic High-Fidelity Music System"

Robert L. Lanier: "The Basic Movements of an Airplane"

James R. Biggs: "Recommendation of a Magnetic Storage Device"

Lindy T. Cergizan: "Recommendation of an Incinerator System"

Len Farr: "The Electric Car and Its Battery"

Thomas Chorba: "Air Bags"

M. A. Spitz: "Recommendation Report on the Purchase of Paint"

Clinton E. Hare, Jr.: "Tests on Concrete for Canyon Project"

Thomas E. McKain: "Feasibility Report on Adjustable-Speed Drive"

Paul Markovich: "Proposal for Protection Against Spread of Alpha Radioactivity"

Robert J. Smith: "Proposal for Installation of a Power Generator"

Robert W. Jonaitis: "Feasibility Report: Site for a Chicago Sport Complex"

William S. Smolen: "Feasibility Report: Leasing or Purchasing a Crane"

Richard K. Hardesty: "Manual on the T-1-A Theodolite Surveying Instrument"

I also wish to thank Mary Jo Shea, Garry T. Atkins, Richard Blastick, Richard Chizmar, Bruce Dunn, Thomas Ference, Robert Gard, Jack Jeka, Marvin F. LaBahn, M. D. McCormack, J. E. McDonald, Jerry G. Morse, Joseph Muszalski, Mike Prospero, Bud Schramlin, James Vater, and Paul Wojtena for ideas or drawings they contributed to this text.

I am grateful to my family, friends, and colleagues for being patient with me during the past two years.

Steven E. Pauley
Purdue University

SECTION ONE

INTRODUCTION

THE TECHNICAL WRITER AND THE REPORT READER

The introductory chapter of this text analyzes the technical writer and the reader of his reports. Most technical writing problems result from people aiming their reports at themselves rather than at others. In your role as a technical writer, you take the first step toward your overall objective, communication, when you fully understand the relationship between yourself and your reader.

1. THE TECHNICAL WRITER AND THE REPORT READER

A REPORT-ORIENTED SOCIETY

"All right. Put it in writing."

These words are heard every day in industry. There, and in society generally, not much counts or becomes official until it is on paper. We have become report oriented.

This seems cold and impersonal, but it is the inevitable result of industry's growth. People just do not have time to spend twenty-five minutes listening to an explanation of how something works, why it went wrong, or how it can be improved. They want the facts in as concise and readable a report as possible. Today the technical specialist usually does not get in the door, but his report does, and it had better do its job while it is there.

It is the technical student's preparation and training that will win him his first job. But his promotion and advancement depend upon his performance, and often on the way he presents himself in writing. Look at it this way: when two men with equal technical ability join the same firm, the one who can communicate will become the other's boss.

Estimates are that the technical man spends at least 25 percent of his time communicating, much of it "writing up" what he has been working on the other 75 percent of the time. It is that 25 percent which has impact. Anyone in industry knows that as a man advances from lower-management echelons on up through middle-management positions, his writing responsibilities steadily increase. Those who promote people recognize this.

BECOMING A BETTER WRITER

No one questions the need for better report writing in business and industry today. Journals and employers in practically all the technical areas lament the failure of technicians and engineers to communicate effectively. The fact that technical writing courses are now mandatory in technical curriculums reflects that need. However, many technical students deny themselves the opportunity to become better writers. To realize their writing potential, they must reject the following rationalizations which so often accompany failure.

A Lot of Great Scientists Are Lousy Writers.   This rationalization is based on the mistaken impression that a scientist who comes up with a great idea automatically becomes a great scientist. In reality, the great scientist is one who comes up with a great idea and communicates it. Albert Einstein had to spell out his complex, revolutionary ideas, and we need only glance at his simple writing style to know that he did a good job of it. It is not inconceivable that had Einstein been a poor communicator people today would say "Albert who?"

This Stuff Is Too Technical to Explain in Simple Words.   Translated, this statement often means, "I don't understand this stuff well enough myself to be able to explain it to you." Anyone who is really proficient in his field can communicate his knowledge to people who have various degrees of technical sophistication. Neil Armstrong can walk into McDonnell Douglas Aircraft in St. Louis and discuss a LEM (lunar extravehicular module) with experts. He can also go on "Meet The Press" and explain the LEM to an uninformed audience. Dr. Paul Ehrlich of Stanford is extremely knowledgeable about ecology and communicated his knowledge to thousands of readers who had never heard of ecology through his book *The Population Bomb.* The late Paul Christman, after his great years in the complex game of pro football, became a broadcaster for television. In eulogizing Christman, Curt Gowdy emphasized his ability to simultaneously put himself in the place of the player and the average viewer. That is precisely the skill necessary for the technical writer. He must be able to talk and write about complex technical subjects, and do it in a way that can be understood by nontechnical people. It is usually not too difficult if the writer thoroughly understands the subject himself.

I'm Good at Technical Subjects, but English Loses Me.   The saddest thing about this rationalization is that if the student is convinced he is a poor writer, his writing will inevitably be poor. There is very little that time, reading, and instruction can do for him. For whatever it is worth, however, it is the experience of most teachers that the student who is good at technical subjects is pretty good at technical writing, too. On the other hand, those who have difficulty with technical writing are often having problems in their other technical courses. There are exceptions, of course, and anyone who chooses to lean on this rationalization is free to do so. However, if you give technical writing equal time with your other subjects, you are likely to find that writing well, although it may not be much fun, beats writing badly.

WRITING HELPS THINKING

Why is it that during the last paragraph of an in-class theme, with three minutes to go, a writer suddenly figures out what he should have been saying all along? Why does writing a rough draft bring us closer to what we want to say? We certainly do not become better writers in the time it takes to scratch together a rough draft. What happens is that we figure out, on paper, exactly what we want to say. The words used for writing are the same ones used for thinking, and through writing we become better, clearer thinkers. The process of getting our thoughts on paper helps bring them into focus, and once they are written instead of floating around in our minds, we can see them clearly and organize our rewriting. We think on paper.

The difficulty we have is often not so much the writing as it is the thinking. If our thoughts are fragmented, indiscriminately spliced, underdeveloped, and disorganized, the odds are that our writing will be all of those things, too. The old definition of a sentence as a complete thought is still accurate. Bad sentences, paragraphs, and reports often result from inadequately thought-out sentences, paragraphs, and reports. Grammar is an easy target, so it usually takes the blame for poor writing, but bad grammar is often a symptom rather than a sickness.

During a technical writing course, students generally become more knowledgeable about the writing topics they choose. Ultimately, they emerge from the course with a better understanding of their technical fields. That is because writing helps thinking.

THE UNINFORMED READER

You will generally write for an uninformed reader. If that seems odd, think of it this way: if he knew more about the subject than you did, he would be doing the writing and you would be the reader.

When asked to write a report in industry, you go through a long process of investigating, thinking, and writing. After you have done all this, you know your subject forward and backward, and when you read the report to yourself, everything is perfectly clear. But you turn it over to a reader who has not been involved in the evolution of the report. The reader gets it cold. Compared to you, he is uninformed about the subject. This does not mean that he is stupid or that things must be repeated for him. It does mean that the report must be written clearly if communication is to be achieved.

ACHIEVING COMMUNICATION

One key to good report writing is using words to communicate with people rather than trying to impress people with them. Anybody can sound impressive without making good sense. Phillip Broughton makes this frighteningly clear with his "Systematic Buzz-phrase Projector." To use it, select any three-digit number; the number, producing a word from each column, will provide a tremendously impressive phrase:

Table 1   *Systematic Buzz-Phrase Projector*

Column 1	*Column 2*	*Column 3*
0. Integrated	0. Management	0. Options
1. Total	1. Organizational	1. Flexibility
2. Systematized	2. Monitored	2. Capability
3. Parallel	3. Reciprocal	3. Mobility
4. Functional	4. Digital	4. Programming
4. Responsive	5. Logistical	5. Concept
6. Optional	6. Transitional	6. Time-phase
7. Synchronized	7. Incremental	7. Projection
8. Compatible	8. Third-generation	8. Hardware
9. Balanced	9. Policy	9. Contingency

Philip S. Broughton, "Criteria for the Evaluation of Printed Matter," *American Journal of Public Health,* 30 (Sept. 1940), 1027–32.

The number 757, for example, produces "synchronized logistical projection," a phrase which could be inserted into any report to impress the reader. As Broughton says, nobody will know what it means but nobody will admit it, either. Industry has no use for this type of writing. The only really impressive writing is writing which communicates.

GUNNING'S FOG INDEX

Robert Gunning created the Fog Index to serve as a practical yardstick for determining the degree of difficulty in any type of writing. It works like this:

One: Jot down the number of words in successive sentences. If the piece is long, you may wish to take several samples of 100 words, spaced evenly through it. If you do, stop the sentence count with the sentence which ends nearest the 100-word total. Divide the total number of words in the passage by the number of sentences. This gives the average sentence length of the passage.

Two: Count the number of words of three syllables or more per 100 words. Don't count the words (1) that are proper names, (2) that are combinations of short easy words (like "bookkeeper" and "manpower"), (3) that are verb forms made three syllables by adding *-ed* or *-es* (like "created" or "trespasses"). This gives you the percentage of hard words in the passage.

Three: To get the Fog Index, total the two factors just counted and multiply by .4.

Using a passage by Albert Einstein, Gunning demonstrates that complex ideas can be expressed in clear, fog-free writing. Let's apply the Fog Index to the passage:

If we ponder over the question as to how the *universe, considered* as a whole, is to be regarded, the first answer that suggests itself to us is surely this: As regards space (and time) the *universe* is *infinite.* There are stars everywhere, so that the *density* of matter, although very *variable* in detail, is nevertheless on the *average* everywhere the same. In other words: However far we might travel through space, we should find everywhere an *attenuated* swarm of fixed stars of *approximately* the same kind of *density.*

Total words: 89
Total sentences (independent clauses count as sentences): 4
Average sentence length: $89 \div 4 = 22.2$
Total three-syllable-or-more words: 10
Percentage of three-syllable-or-more words: $10 \div 89 = 11.2$

$22.2 + 11.2 = 33.4$
$33.4 \times 0.4 = 13.3$ *Fog Index*

To clarify what the 13.3 means, let's again turn to Gunning, who compares the Fog Index with the reading levels required for various magazines and grades in school:

Table 2   *Reading Level*

	Fog Index	By Grade		By Magazine
	17	College	graduate	
	16	"	senior	(No popular
	15	"	junior	magazine
	14	"	sophomore	this difficult.)
Danger				
Line	13	"	freshman	
	12	High-school	senior	*Atlantic Monthly* and *Harper's*
	11	"	junior	*Time* and *Newsweek*
	10	"	sophomore	*Reader's Digest*
Easy-	9	"	freshman	*Saturday Evening Post*
reading				
Range	8	Eighth	grade	*Ladies' Home Journal*
	7	Seventh	"	*True Confessions* and *Modern Romances*
	6	Sixth	"	Comics

Robert Gunning, *The Technique of Clear Writing*, rev. ed. (New York: McGraw-Hill, 1968), p. 40.

The table indicates the reading levels at which popular magazines are written. They seem to be aimed very low, but of course their circulation depends upon reaching uninformed readers. The only exception to Gunning's statement that no popular magazine's Fog Index is above twelve is *Playboy*, which might be evidence that more people look at it than read it. According to the table, Einstein's passage could be understood by a high school junior, which is amazing in view of his subject matter. Among his theories must have been one which said "keep writing simple."

There is no excuse for making written communication more complex than the topic requires. Even if the information is highly technical, the writer must communicate it as clearly as possible. A nation-wide survey by Richie R. Ward, author of *Practical Technical Writing*, showed that even highly educated and experienced technical men prefer reading reports that are below the Fog Index of fifteen. To make the right decisions, readers need to thoroughly understand the information contained in reports.

THE VALUE OF VISUALS

Visuals can greatly assist your effort to communicate technical information. Most people are more adept at grasping visual information than information presented in words. Technical visuals, including photographs, drawings, charts, diagrams, and tables, make reports more attractive; and, more important, they reinforce the paragraphs of the report. You can select the type of visual that corresponds to your reader's level of technical knowledge. Chapter 7 provides a detailed explanation of how to use visuals effectively.

Too many reports are written in a vacuum. Writers write for themselves instead of aiming at an uninformed reader, and then wonder why their reports fail to communicate. Become conscious of your reader, and try to write reports which defy misunderstanding. From the reader's point of view, the only good technical writing is fail-safe writing.

SECTION TWO

TECHNICAL WRITING TECHNIQUES

DEFINING / DESCRIBING A MECHANISM /
DESCRIBING A MECHANISM IN OPERATION /
INTERPRETING STATISTICS /
RESEARCHING PUBLISHED INFORMATION /
ILLUSTRATING

SECTION TWO OVERVIEW

This section of the text describes the basic techniques which you will need for writing in industry. Practically every word in your reports will fall into one of these categories: definition, description, or interpretation of statistical data. Your descriptions will not always be limited to mechanisms, of course, but the principles for describing mechanisms and their operations will apply to your reports in industry. Research techniques will prepare you for other academic assignments as well as for keeping aware of advancements in your field after you graduate. Illustrating is not specifically a writing technique, but it does enable you to reinforce the words in your reports.

2. DEFINING

Definition is a logical technique by which the meaning of a term is revealed. It has become important in this era of technical specialization because each technology has developed words which are not familiar to people outside that technology. Technicians in industrial engineering, for example, cannot expect a specialist in metallurgy or civil engineering to be aware of their terminology. And it is a rare mechanical engineering technologist who can wander into an electronics laboratory, spend five minutes listening to words like *heterodyne detection, pyrometer,* and *micromho,* and emerge with more than a vague idea of what is going on. Such specialized terminology is appropriate for informal reports within a department, but not for reports going beyond the department. This communication technique becomes critical in large industrial firms, where specialists in various technical areas must function together and communicate with each other in a coordinated effort toward the firm's objectives.

Even if communication among various technologies presented no difficulty, the problem of communication with management would still exist. Just as the number of technologies has multiplied, diversification has occurred in management, and specialists are required to administer particular areas. Although some management personnel have come up through the ranks, and many others have strong technical backgrounds, they cannot realistically be expected to keep up with various technological advances while they function as administrators.

In short, the era of technical generalists, individuals with working knowledge of several technologies, has passed. People have enough trouble keeping up with innovations in their own fields. Report writers today must recognize this and write accordingly. Abbreviations and symbols, which were once used to save space, are now omitted from reports because the space saved is not worth the communication lost. Words are carefully chosen and defined because there is no excuse for making the reader look up a word or ask for clarification. Neither is there any reason for allowing a reader to think he knows what a word means, only to discover later that the intended meaning was entirely different. Reports are written to save time rather than waste it.

INFORMAL DEFINITIONS

Definitions should be avoided whenever possible; if a simple word conveys the intended meaning, use it. However, when technical terminology cannot be avoided, an informal, conversational definition is often sufficient. Though less thorough than a formal definition, the following types of informal definition are quite acceptable if they furnish needed information to the reader.

OPERATIONAL

Operational definitions describe an activity which produces the effect being defined. For example, "frictional electricity," an effect, can be illustrated by describing how a person becomes negatively charged by the friction between his shoes and a rug; when he touches an object, the excess electrons rush off, causing a spark and a small shock. A word like *acceleration* can be partially explained in terms of what happens when a driver steps on the gas pedal of his automobile.

NEGATIVE STATEMENT

Negative statements explain what a term does not mean, often in order to correct a common misconception. To be effective, a negative statement must be followed by a positive one. You might, for instance, say that a "funnycar" is not necessarily a funny-looking car. You would then explain that "funny" derives from "phoney," which refers to the car's fiberglass rather than metal body. A series of negative statements can perform a process of elimination leading to the right definition. You might explain that a construction technologist cannot properly be called either a draftsman or an architect, and then make distinctions among the three to arrive at an accurate, positive definition.

Antonyms, or words having the opposite meaning, can be useful negative statements also, but again the reader should not be expected to fully understand a term if merely informed of its opposite. A positive statement must follow.

SYNONYM

Synonyms are effective only when the word used is more popular than the term being defined. People are more familiar with *voltage* than *electromotive force,* the technical term carrying the same meaning. Other examples are *spiral,* for *helix,* and *spun glass* for *fiberglass.* When aiming at an uninformed reader, never pass up an opportunity to clarify a technical term with a more common word, and if possible, avoid the difficult word entirely by merely substituting the easier one.

In daily conversation, we begin informal definitions by saying, "What I mean is . . . ," or, "Look at it this way." The same conversational approach can be used to clarify words and ideas in technical reports. Nothing is wrong with writing "In other words . . . ," or "In this report, *environment* refers to . . . ," or simply placing a synonym in parentheses. Clear and conversational writing communicates with the reader because it presents technical information in his own language.

FORMAL DEFINITIONS

Every technology has its precision instruments. Learning how to use them requires patience, but they ultimately make your job easier. In technical writing, the formal sentence definition functions like one of those instruments. It "clears the board" by scientifically separating the thing being discussed from everything else in the world. The definition's three parts are the item (species) which needs defining, the class (genus) to which the item belongs, and the differentiation (differentia) of the item from all other members of its class:

Item	*Class*	*Differentia*
acceleration	the rate of change	of velocity with respect to time
resistance	any force	that tends to oppose or retard motion
dynamics	the study	of the relationship between motion and the forces affecting motion

Item	*Class*	*Differentia*
glider	a light, engineless aircraft	with lift surfaces and extended wings, designed for long periods of flight after launch from a towing vehicle
howitzer	a cannon	with a barrel longer than a mortar that delivers shells in a high trajectory against targets that cannot be reached by flat trajectories
engineering	the application	of scientific principles to practical ends such as the design, construction, and operation of efficient and economical structures, equipment, and systems
ammeter	an instrument	that measures electric current
technique	a systematic procedure	by which a complex or scientific task is accomplished
electricity	a physical phenomenon	arising from the existence and interactions of charged particles
carburetor	a mixing chamber	used in gasoline engines to produce an efficient explosive vapor of fuel and air

CLASSIFYING THE ITEM

As shown in the preceding examples, careful modification of the class goes a long way toward completing the definition. The narrower the class, the more meaning it conveys, and the less that needs to be said in the differentia. Classifying an item as an "intricate device" accomplishes very little; the world is full of intricate devices, and the item still needs to be differentiated from all of them. Also, saying that an item "is what" or "occurs when" suggests that those are words of classification, and they obviously are not.

Such words can be useful in informal definitions, but have no place in formal sentence definitions.

DIFFERENTIATING THE ITEM

If the differentia applies to more than one item, the definition lacks precision. Even simple items like ballpoint pens require cautious differentiation, for the pens are distinguishable only by the design of their points. You must also guard against circular definitions, which are caused by the repetition of the very word you are defining. The reader who needs help with *capacitance* will get nowhere if expected to understand the word *capacitor* in the differentia. Noncrucial words such as *writing* in the item *technical writing* can, of course, be repeated. Occasionally, an extensive differentia requires a second sentence.

AMPLIFYING FORMAL DEFINITIONS

After reading a formal sentence definition, an uninformed reader often needs further explanation to really understand the item. Seven methods for amplifying definitions are discussed below.

EXPLICATION

In this context, explication means defining difficult words contained in the formal sentence definition. The words *velocity* and *charged particles* in the examples above would need defining for many readers. When explicating, you can often write an informal definition rather than another formal one.

DERIVATION

The origin of some words helps clarify their meaning. *Ammeter,* for example, derives from the words *ampere* and *meter,* and *scuba* is made up of the first letters of the words *self-contained underwater-breathing apparatus.*

COMPARISON

Never pass up an opportunity to compare the item being defined to a similar item that the reader is familiar with. In defining a howitzer, you might emphasize its likeness to a cannon. The *Supervi-*

sor's Handbook for U.S. Steel Supply draws upon the readers' knowledge of golf to clarify a term: "The Standard Cost System is a control for comparing what we actually spend (the strokes we took) with what we should have spent (par)."

ILLUSTRATION

Drawings and diagrams are very effective tools for reinforcing definitions. A small, well-labeled drawing would help explain the function of a carburetor, and "resistance" could be amplified by a diagrammatic representation of forces in opposition.

EXAMPLE

The best examples are simple ones, as shown in the following discussion of *resistance:* "If a boy runs through rooms with the doors open, he experiences very little resistance. If one of the doors is closed, resistance occurs. The boy must stop and open the door or run right through it. Either alternative will produce resistance to his movement." Effective examples also take advantage of the reader's knowledge: a good opportunity would be missed if an amplification of *weightlessness* did not refer to our astronauts' experience in space.

ANALYSIS

Analysis refers to the division of an item into its main parts. This method aids the reader's comprehension by allowing him to grasp the definition bit by bit. For example, *dynamics* becomes easier to understand when its two main parts, kinetics and kinematics, are discussed individually.

CAUSE AND EFFECT

Some phenomena are so elusive that they must be clarified in terms of their causes and effects. An amplified definition of *electricity* would probably explain the interactions that produce it, and *magnetism* might be approached through a description of the force it produces.

Successful amplified definitions generally combine several methods of amplification, the most common being analysis and example.

When writing an extended definition, you will immediately identify a couple of applicable methods, and using them will lead you to other methods.

On the following pages are four student papers which will serve as models. The first two papers use analyses and examples to amplify the formal sentence definitions. The third one uses an example, and the fourth model relies on an extended example which gradually becomes more complex to reach the uninformed reader. Examine the models critically, and offer your class suggestions for improving them.

STRESS AND STRAIN

In engineering, stress and strain describe the conditions of a material under varying circumstances. Stress is the material's resistance to external forces, measured in terms of the force exerted per unit of area, and strain is the material's change resulting from those external forces. Every force that acts on a material creates a type of stress, and each stress has an accompanying type of strain. The three basic types of stress are tensile, compressive, and shearing.

Tensile stress occurs when a pair of forces act on opposite ends of the material and attempt to pull it apart. The resistance the material offers to this action is known as the tensile stress in the material. Tensile stress is most easily demonstrated by pulling on the ends of a rubber band. The type of stretching (strain) taking place in the rubber band occurs to some degree in all other materials which are under this type of force.

Compressive stress results from forces pushing in on the ends of a material. A simple example of a compressive force is pressure exerted to push an accordian together. An accordian has only a small compressive stress because the accordian bag offers little resistance to being forced together. Reduction in the length of the material (strain) accompanies a compressive stress.

Shearing stress occurs when a force acts downward on a material, causing a piece of it to bend or break off. This type of force, known as shearing force, is resisted by shearing stress. Shearing force and stress can be demonstrated by slicing a loaf of bread. Although amplified by the knife's cutting edge, the force can still be considered a shearing force. Like all other materials, the bread offers resistance (stress) but tends to bend or break (strain) under such a force.

DYNAMICS

Dynamics is the study of bodies in motion and the effect of forces acting on bodies in motion. Dynamics can be broken into two areas, kinematics and kinetics.

Kinematics examines the motion of bodies without consideration of their mass (weight) or the forces causing the motion. One can observe the distance a body moves during a given time period and, by the use of formulas, predict where the body will be at a later time. This type of dynamics is very useful in the study of free-falling projectiles. When the guns on a battleship are fired, it is possible to predict with extreme accuracy how far the projectiles will go, and how long they will stay in flight. In civil engineering, kinematics assists in the design of curves in our highways; the engineer uses certain frictional values between the tires and the pavement to find a degree of embankment that can be safely taken at 70 miles per hour. The forces which cause the motion of the projectile and the automobile are not examined in kinematics.

Kinetics, on the other hand, considers the forces acting on bodies in motion. For example, an automobile going slowly around a corner possesses less inertia (the property of a body that resists change in speed or direction) than the same automobile rounding the same curve at a greater speed. The forces caused by the different velocities must

be evaluated even though the automobiles weigh the same.
The design of an aircraft also involves kinetic problems.
The net wing area necessary to lift the 747 Jumbo Jet was
computed through use of kinetic formulas, and so was the
amount of thrust necessary to give the wings this potential.
Action-reaction forces (Newton's third law: for every action
there is an equal and opposite reaction) must be evaluated
in order to get the plane off the ground. Thus, kinetics
goes one step beyond kinematics because it deals with all
aspects of motion, including kinematics.

AUDIO-VISUAL SYNAESTHESIA

Audio-visual synaesthesia is a perceptual effect which simultaneously stimulates the senses of vision and hearing. Each sound sensation is perceived as harmoniously interdependent with a visual sensation.

Synaesthesia is an art form which stems directly from technological roots. The advances in electronics have made possible the stimulation of more than one sense at a time, and future technological advances may allow the creation of a total perceptual environment.

One of the simplest devices for demonstrating the effect of audio-visual synaesthesia is the color organ. A color organ is an electronic circuit which converts the varying beats and intensities of recorded music to a corresponding light display. The lights flash with the beat of the music, and their brightness varies with the volume. Some color organs have circuitry which displays the high tones of a musical passage as one color, the medium tones as another color, and the low tones as yet another. The overall effect of watching the display while simultaneously listening to the music is that one can sense that he is "seeing" the music, or "hearing" the visual display.

In audio-visual synaesthesia, two usually separate sensations are fused into one harmonious perceptual effect. Synaesthesia may be the art form of the future.

CHEMICAL EQUILIBRIUM

"Chemical equilibrium" refers to a unique type of reaction in which the rate of formation of a product equals
the rate at which the product decomposes. All chemical re-
actions are, to some extent, reversible, and to explain the
quality of reversibility, a chemical equation must be examined.

The group of symbols in a chemical equation represents
elements (O-oxygen, H-hydrogen, S-sulfur, etc.) and the nature of the product that can be obtained from combining the
elements. Much as the algebraic equation has two sides
separated by an equal sign, the chemical equation has a reactant side and a product side. The reactant side represents the elements or compounds (two or more elements chem-
ically combined) started with, and the product side represents the elements or compounds remaining after reaction.
The two fractions of such an equation are separated by a
modified equal sign known as a yield sign ($\longrightarrow$).

The following is a simple equation depicting the formation of water from the elements of oxygen and hydrogen, O
and H:

$$2H_2 + O_2 \longrightarrow 2H_2O$$

The symbols "$2H_2 + O_2$" constitute the reactant side, and the
$2H_2O$ constitutes the product side. The yield sign indicates
that the reactant and product sides may be interchanged

without affecting the validity of the equation. One can
therefore begin with water and end up with its components,
hydrogen and oxygen; that is, water can be decomposed, or
broken down. Symbolically, this appears as follows:

$$2H_2O \longrightarrow 2H_2 + O_2$$

To restate the definition of chemical equilibrium in
terms of the water equations, one can say that the rate at
which the hydrogen combines with oxygen to form water equals
the rate at which the water reverts back to its separate components of hydrogen and oxygen. The reaction is in equilibrium.

WRITING ASSIGNMENT

Write a 250-word amplified definition of a term (not an object) from your technical area. Begin with a formal sentence definition, and use the methods of amplification which will make the word meaningful for an uninformed reader. Try to select a term which does not require you to look up the formal sentence definition. If you use other sources, however, give credit to them.

In industry, you will probably never have to write a definition of this length, but mastery of the technique will prove useful in all reports. Some terms which might be narrowed to your field are:

fail-safe	elasticity
hypothesis	thrust
compression	objectivity
feedback	lift
potential	functional form

EXERCISES

1. Write formal sentence definitions of the terms below, assuming that your reader has no familiarity with them. As you classify each item, avoid saying "is what" because those are not words of classification. Also, narrow the class as much as possible by using modifiers; this will reduce the number of words necessary in the differentia. Construct your differentia carefully: the function of a formal sentence definition is to distinguish an item from every other item in the world.

profit	radar
algebra	revolutions per minute
quarterback	hockey puck
horsepower	median
kite	friction

2. Select one of the items above and write three paragraphs amplifying your formal sentence definition. The possible methods of amplification are explication, derivation, comparison, example,

illustration (visual), analysis, and cause and effect. Make sure that each of the paragraphs has a topic sentence stating its central idea, and that your writing is aimed at an uninformed reader.

3. Working with other members of the class who are majoring in your technical area, select three terms which are commonly used and understood in your classrooms or laboratories. As a group, write formal sentence definitions of the terms. Then, ask the other groups to provide oral definitions of the terms. The purpose of this experiment is to determine whether people in other technologies are informed about terminology used in your technical area.

4. In class discussion, compare the following amplified definition of "stress" to the definition contained in "Stress and Strain" earlier in the chapter. Ignore the "strain" portion of the earlier paper and concentrate on evaluating the two papers' explanations of "stress." What methods are used to amplify the formal sentence definitions? Are the methods used effectively? Should other methods of amplification be included? Are statements in either paper unclear, misleading, or inaccurate? What items would you include in an amplified definition of "stress?"

STRESS

"Stress," as defined in engineering, is any resistance to external forces acting on a member. Stress is measured by the force per unit of area. Forces are expressed in pounds per square inch, so stress is produced in any member upon which external forces are acting. The three basic types of stress are (1) tensile stress, (2) compressive stress, and (3) shearing stress.

Tensile stress is created when a pair of axial forces pull on opposite ends of a member. The result of this action is that the member tends to stretch or elongate. This is known as tensile stress and is illustrated in Figure 1.

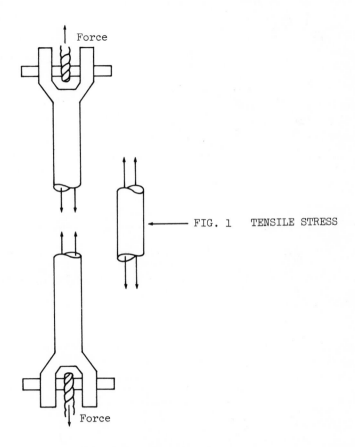

FIG. 1   TENSILE STRESS

Compressive stress is produced when a pair of axial forces push on opposite ends of a member, attempting to smash or crush it. This action, known as compressive stress, is illustrated in Figure 2.

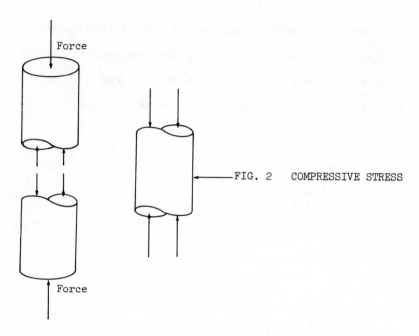

Force

Force

—FIG. 2   COMPRESSIVE STRESS

The last stress is shearing stress, which occurs when a force acting downward bends or breaks off part of the member. This can be illustrated by a bolt being sheared by excessive loading (Figure 3).

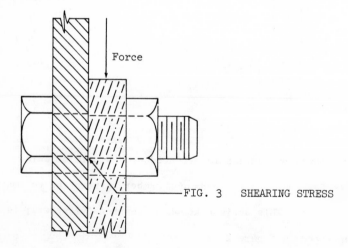

Force

—FIG. 3   SHEARING STRESS

3. DESCRIBING A MECHANISM

Description is more widely used than any other technical writing technique. In fact, you will have to do some describing in every report you write. A mechanism's physical characteristics must be understood before its operation can be grasped. What size is it? What material? How is it powered? Can it take a beating? These and a hundred other questions require answers.

Description is most frequently used in manuals, where the writer provides a detailed physical description of the mechanism before explaining how it is operated. In addition to manuals, however, descriptive sections are necessary in proposals, feasibility reports, and short, informal reports. The feasibility writer, who has been asked to report on several alternative systems, describes and compares them before recommending one for installation. In proposals, the writer tries to sell his firm's design by providing the potential buyer with a thorough description of it. These reports cannot contain vague or ambiguous writing. The descriptions must be detailed, clear, and accurate to make the reader feel secure in his knowledge of the mechanism.

THE ALL-INCLUSIVE MECHANISM

A mechanism is any system of parts that operates in a definable way. In other words, not only a machine but almost anything that has a specific function can properly be called a mechanism. The following description of an airplane hangar demonstrates the inclusiveness of the word *mechanism*. The quotation is from a *Design News* article (Feb. 3, 1969) entitled "Hyperbolic Paraboloid of Steel Forms" by Lars Soderholm, and its purpose is to describe the newly designed hangar to people in the construction industry:

The hangar is supported by front and rear fascias built in the form of canted A-frames with their legs resting on huge concrete buttresses. Longitudinal tension bars are draped between the fascias and serve to support the roof almost in the same manner as cables from a suspension bridge. Wide-flange compression struts or purlins run diagonally between the fascias to prevent them from bending inward. They also provide vertical support for the roof decking and snow loads.

The front fascia, constructed of wide-flange beams, holds the nose section of the aircraft. The rear fascia is made up of a truss section and will cover two king-sized, self-propelled doors that swing open on a steel track, offering a 164-ft. opening. (Pp. 26–27)

The article is evidence that each technical discipline in our highly technical society contains items which lend themselves to description. Readers in all fields need to know how things are constructed in order to clearly understand how they work. For this reason, the word *mechanism* will be used comprehensively throughout this chapter and Chapter 4. The principles of description can usefully be applied to such diverse items as a fuel-injection system, diode, arch, bridge, mechanical pencil, gate valve, and innumerable other mechanisms in all technical areas.

REACHING THE READER

A mechanism must be described as carefully as possible to achieve communication with an uninformed reader. First, you have to be thoroughly familiar with the mechanism yourself. If possible, get your hands on it, take it apart, and examine it closely. Most of the battle is won if you know the mechanism inside out. On the other hand, you can become a victim of your own knowledge. Things that seem obvious, but are crucial to the reader's understanding, can easily be overlooked. The reader does not have a chance if you accidentally omit an essential fact.

Another way of reaching the uninformed reader is through the use of visuals. The *Design News* article quoted earlier was backed up by several diagrams and photographs. A visual can reinforce the description of a complex mechanism—and the key word is *reinforce*. Visuals aid or supplement the description, but they do not speak for themselves; they should never stand alone. Also, using a visual simply because it looks good is as defeating as using words that only look good. Visuals should be reserved for occasions when they will assist the effort to communicate. The type of visuals you select depends upon the mechanism and the reader. A cross-sectional drawing may be better than an exploded view for clarifying a particular piece of equipment. Your options concerning visual material are discussed in detail in Chapter 7.

OUTLINE FOR DESCRIPTION OF A MECHANISM

Guidelines for describing a mechanism are presented in the outline below. The word *Introduction* appears as a main heading in the report, but the word *Description* does not. The main headings in the report's description section will be the names of the main parts. A detailed explanation of what is required in the introduction and body immediately follows the outline.

I. Introduction
   A. Definition and purpose of mechanism
   B. Overall appearance
   C. Identification of main parts

II. Description
   A. Part (Component) One
      1. Definition and purpose
      2. Appearance
         a. Size
         b. Shape
         c. Material
         d. Location and method of attachment
   B, C, etc.—Description of other parts

THE INTRODUCTION

In the introductory section, provide an overview by defining the mechanism and its purpose, describing the mechanism's general appearance, and naming its main parts. An effective introduction sends the reader into the body of the paper, the part-by-part description, with enough general information for him to be able to grasp the specifics of the mechanism.

DEFINITION AND PURPOSE

The uninformed reader needs a formal sentence definition of the type discussed in Chapter 2. For a reader who is familiar with the mechanism's general class, a more informal definition, emphasizing the differentia, may be sufficient.

Sometimes the definition adequately explains what the mechanism does, but an explicit statement of purpose often requires an additional sentence. For example, to say that an air conditioner cools a room is to give the reader little information; a breeze through an open window does much the same thing. A clear explanation of an air conditioner's purpose involves such things as circulation, filtration, and dehumidification.

OVERALL APPEARANCE

In this section, state the dimensions and shape of the entire mechanism to give the reader a point of departure. Along with the statement, a drawing with the main parts labeled can assist both an informed and uninformed reader, as a glance at any good manual will indicate. If an illustration cannot be included, help the reader by comparing the overall mechanism to something he is better acquainted with. For example, pointing out the similarity of a dune buggy to a jeep takes advantage of the reader's familiarity with the more common mechanism.

IDENTIFICATION OF MAIN PARTS

The final portion of the introduction identifies the main parts of the mechanism. The part-by-part description in the body of the report will be organized according to those main parts, with each receiving a heading. This approach, which applies to any type of technical report, allows the reader to assimilate the description bit by bit. Fortunately, all mechanisms lend themselves to logical analysis, so breaking them into their main parts causes few problems. For example, if you were describing a pencil sharpener, you would probably work from exterior to interior: the first main part would be .the casing, which attaches to the frame; next, the frame; then, the cranking mechanism, which extends into the mechanism's interior; and finally, the gear mechanism, which rests entirely in the interior.

As you identify the mechanism's main parts, list them in the order they will be described in the body. This short statement at the end of the introduction tells the reader exactly what to expect in the rest of the report.

DESCRIPTION

In outline form, the detailed description of a part looks difficult, but once you actually start describing, things will fall into place. A single paragraph usually suffices for each part. Use the "Definition and purpose" of the part, stated informally, as the topic sentence for each paragraph. When necessary, include a drawing of a complex part to reinforce its detailed description.

APPEARANCE

After you have written a topic sentence, you must consider five elements: size, shape, material, location, and method of attachment. The following description of a gear system, one of the four main parts of a pencil sharpener, contains all of them. Although the system has three subparts, they are not difficult to fit into a coherent paragraph:

```
    The gear system consists mainly of three gears and two

cylinders, all made of steel. The two identical cylinders

are 1 inch in diameter and 1-1/2 inches long (Fig. 1).  Each

has 17 spiral edges which perform the cutting operation on

the pencil.  In order to sharpen the pencil to a 15-degree

cone, each cylinder is held at a 7-1/2 degree angle by a shaft

which is screwed through the cranking mechanism (not shown).

At one end of the cylinders are small gears with 9

teeth each.  These teeth mesh with 22 teeth inside a station-

ary main gear 1-3/8 inches in diameter which is welded to a

frame.  The three gears form what is called a planetary gear

system.
```

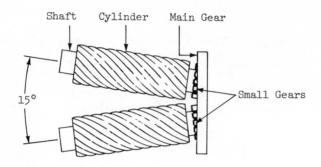

Shaft   Cylinder   Main Gear

15°

Small Gears

FIG. 1   GEAR MECHANISM

SIZE

In the example above, the sizes of the three main gears are stated within the text. If several more dimensions were needed, the smaller ones could be placed on the drawing to avoid cluttering the text, which would then include only overall dimensions. Depending on the purpose of the description, writers in industry provide either extremely thorough dimensions or, as in the example above, just enough dimensions to give the reader a perspective of the part. The use of numbers and hyphens in the example conforms to generally accepted guidelines which are discussed in "Common Technical Writing Errors," the Appendix of this text.

SHAPE

In the description of the pencil sharpener's gear system, the words *cylinder* and *gear*, which indicate roundness, clarify the shape of the parts. Many times a simple analogy, comparing the part to a letter of the alphabet (A, C, L, T, U, among others) adequately suggests its shape.

MATERIAL

All the subparts of the gear system are made of the same material (steel). If an entire mechanism consists of only one material, you can state this in the introduction to avoid bothering with it in the detailed description.

LOCATION AND METHOD OF ATTACHMENT

Location and method of attachment refer to the position of the subparts in relation to each other. To clarify these relationships, you must examine the main part and determine a logical approach to describing its subparts. The gear system's subparts are described in an order which reinforces their relationship to each other. Working essentially from left to right, the writer describes cylinders and their position in reference to each other, states how they connect to the cranking mechanism, how the gears mesh with the main gear, and how the main gear attaches to the frame.

After describing the first main part of a mechanism, follow the same procedure for the remaining parts. A mechanism is the sum of its parts, and detailed descriptions of each add up to a detailed description of the entire mechanism. Descriptions of this type generally require no conclusion; a well-written description leaves nothing to conclude.

Remember that description is one of the most important techniques of technical writing. In industry, you will never write a report that does not require some description, and you may occasionally write one that contains nothing but description.

As you read the four student models which follow, observe their organization. Your analysis of a mechanism and the order in which you describe the subparts of each main part will have tremendous bearing upon how well you communicate. Also, offer your class specific suggestions for improving the models. For example, the "Description of an Automatic Center Punch" has several strong points, but is the description repetitious? Are all the parts adequately described? How would you improve the description?

DESCRIPTION OF AN AUTOMATIC CENTER PUNCH

INTRODUCTION

The automatic center punch is a device used to make a
pinpoint indentation in a hard material such as metal or
plastic without the use of a hammer. The indentation provides a more stable starting point for a drill to cut the
material. The automatic center punch (Fig. 1) resembles a
short screwdriver, but instead of having a tip like a screwdriver, it has a round metal shaft sharpened to a conical
point. Four basic parts make up the center punch (Fig. 2):

1. the handle, containing a driver assembly which consists
    of a very stiff spring and a releasing member;

FIG. 1   ASSEMBLED VIEW

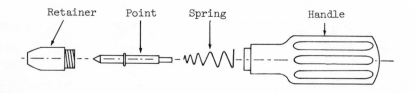

FIG. 2   EXPLODED VIEW

2. the return spring;

3. the point, sharpened to a conical shape;

4. the retainer, which keeps the point, the return spring,
   and the handle together.

HANDLE

The handle, which resembles that of a screwdriver, is
approximately 2-1/2 inches in length and 13/16 inch in diameter.
The handle's plastic exterior permanently encases a stiff
spring and a releasing member. The spring is the part that
drives the point to the material.

RETURN SPRING

The return spring is a steel spring not much different
from the type used in most ballpoint pens. It returns the
point to its normal position after the tool has been used.

RETAINER

The retainer is a tubular piece of steel which contains
the return spring and the point. Its inside diameter is
17/64 inch at its wide end and 5/32 inch at its narrow end.
The retainer is 43/64 inch from end to end and has 32 threads
per inch at its wide end, which screws into the base of the
handle.

POINT

The point is a steel bar 1-5/16 inches long which is machined and polished to a main diameter of 0.1532 inch. The end opposite the point measures 0.0923 inch in diameter for a distance of 3/8 inch. Twenty-five thirty-seconds of an inch from this same end, a groove is cut completely around the shaft. A separate piece of wire wraps around this groove to hold back the return spring. The pointed end of this shaft is sharpened to a 60-degree angle and hardened to resist wear.

DESCRIPTION OF A ONE-INCH MICROMETER

INTRODUCTION

A one-inch micrometer (Fig. 1) is a hand device for
making outside-diameter measurements from 0.000 inch to
1.000 inch. Its basic purpose is to measure small objects
which require accuracy up to 0.001 inch.

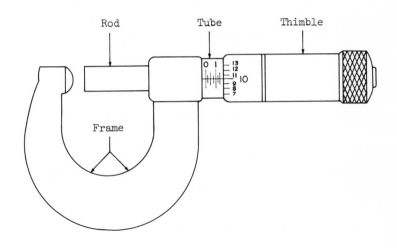

FIG. 1   ONE-INCH MICROMETER

The main parts are the frame, a 3/8-inch tube fastened
to the frame, a 1/4-inch finely machined rod, and what is
called a thimble attached to the rod's threaded portion by a
small screw.

FRAME

The frame is made of stainless steel and machined to a
fine finish. The head, which also has a stainless, finely
finished surface, mounts on the frame (Fig. 2).

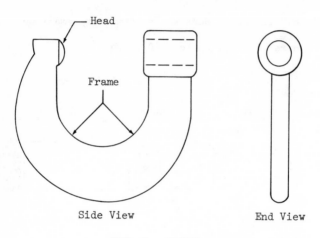

Side View   End View

FIG. 2   FRAME

TUBE

The 0.375-inch-diameter stainless steel tube (Fig. 3)
attaches to the frame. A 0.250-inch-diameter hole runs
through the tube and the frame, allowing insertion of a rod
through the tube. Internal threads (46 per inch) on the inside of the tube are very important to the accuracy of the
micrometer and will be discussed with the threaded portion of
the rod. One inch of the tube is marked to indicate divi-
sions of 0.025 of an inch. The 0.100, 0.200 and 0.300, etc.,
are marked by 1, 2, 3, etc., to zero, which indicates 1.000
inch.

-2-

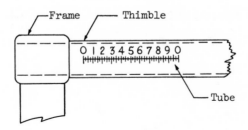

FIG. 3   TUBE

ROD

The rod, made of finely machined stainless steel, has threads on the back half to match the internal threads of the tube. The threads provide the micrometer's great accuracy. A complete turn of the rod equals 0.025 of an inch toward or away from the head on the frame, depending on the direction it is turned (Fig. 4).

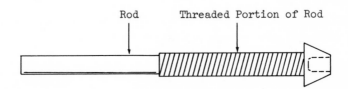

FIG. 4   THE ROD

THIMBLE

A stainless steel thimble (Fig. 5) is attached over the threaded portion of the rod by a small screw. Markings divide the thimble's circumference into 25 increments, each

-3-

of them representing 0.001 of an inch. The thimble has a
knurled finish on the end for easy handling with the thumb
and index finger.

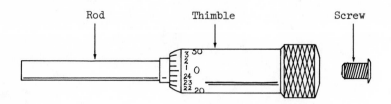

FIG. 5   THIMBLE

DESCRIPTION OF A MOVING-COIL METER

INTRODUCTION

The moving-coil meter is an electric-current-indicating device. As its name suggests, this instrument relies upon a moving coil of wire to indicate the presence of an electric current. The device is often housed in a case with a calibrated scale and used as an ammeter, voltmeter, or ohmmeter. Other names for the moving-coil meter are galvonometer,

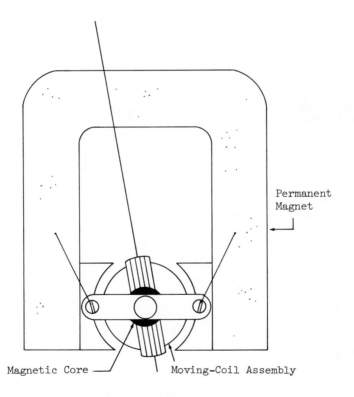

Permanent
Magnet

Magnetic Core ——   Moving-Coil Assembly

FIG. 1   MOVING-COIL METER

-1-

Weston-movement, or D'Arsonval-movement, all of which are
various forms of the moving-coil meter.

The meter is an electromagnetic device which consists
of the following basic parts (Fig. 1): a permanent magnet, a
moving-coil assembly, and a magnetic core.

PERMANENT MAGNET AND POLE PIECES

The horseshoe-shaped permanent magnet (Fig. 2) comprises
the major part of the moving-coil's magnetic circuit. It is
constructed of hard steel, usually alnico. This hard steel
retains a high percentage of magnetism for a great length of
time, which is very desirable.

The pole pieces are made of soft iron and attached to
the horseshoe-shaped magnet. The pole pieces detour the magnetic circuit's path and concentrate the magnetic flux in
the area where the moving-coil is located. Magnetic flux in

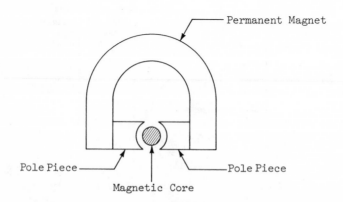

FIG. 2   SIMPLIFIED DIAGRAM OF PERMANENT
MAGNET AND POLE PIECES

a magnetic circuit is similar to electric current in an electric circuit.

MOVING-COIL ASSEMBLY

The moving-coil assembly is composed of an aluminum frame (Fig. 3) which is the supporting structure for the coil. The moving-coil mechanism is mounted on pivots, and the ends of the pivots are set into jeweled bearings. The needle attaches to the moving coil and indicates the position of the coil on a scale. Springs, which are located near the pivot points, supply torque to oppose the force developed by the current flowing in the coil and allow a smooth deflection of the needle. The torque springs also connect the coil to terminals, located on the meter housing, for external circuit connections.

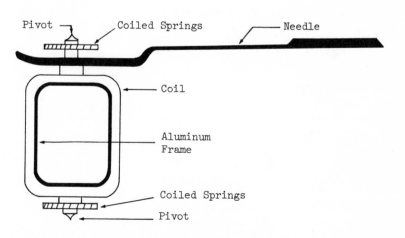

FIG. 3   MOVING-COIL ASSEMBLY

MAGNETIC CORE AND MOUNT

The magnetic core and mount (Fig. 4) are located in the center of the moving-coil structure and reinforce the magnetic circuit that envelops the moving coil. The core is made of soft iron, like the pole pieces, and helps smooth the moving coil's movement.

The mount for the magnetic core holds the core in position. It also provides the mounting points for the jeweled bearings which support the entire moving-coil assembly.

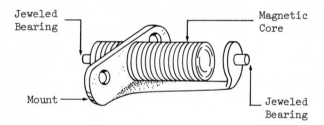

FIG. 4   MAGNETIC CORE AND MOUNT

DESCRIPTION OF A LONG-NOSED PLIERS

INTRODUCTION

Long-nosed pliers are designed for use in limited
operating space. In the electronics industry, they are
widely used to reach out-of-the-way terminals and connectors.
The pliers are approximately 6 inches long and composed of
two halves that are mirror copies of each other (Fig. 1). A
specially designed rivet fastens the two halves together.

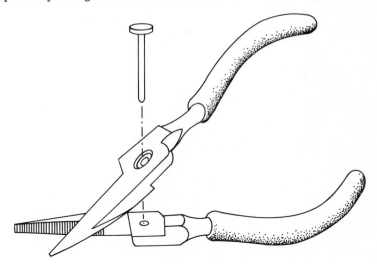

FIG. 1.   LONG-NOSED PLIERS

THE TWO HALVES

The two halves of the pliers are "drop-forged," which
is a process of forming metal by dropping a high impact
hammer on hot metal, forcing it into a mold. The handles of
the pliers are curved to fit the hand, and have a plastic
covering which acts as a grip and helps insulate the user

from electrical shocks. The center of the pliers widens from the handles to allow room for the rivet which fastens the two halves together. The nose of the pliers tapers down to the tip, where the jaws are quite thin. The stronger drop-forged metal allows the thin jaws to take much abuse without bending or breaking. Each jaw has small teeth enabling the pliers to grasp objects. Teeth and the thin, tapered nose make these pliers invaluable in tight spots where ordinary pliers cannot be used.

THE RIVET

A rivet connects the two halves. It has a flat head on one side that fits into a countersunk hole the same size as the rivet head (Fig. 2). The rivet's shaft fits through previously drilled holes in the two halves. To hold the two halves together, the portion of the shaft that extends beyond the bottom half must be flattened out into a countersunk hole.

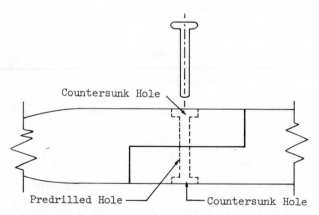

FIG. 2   RIVET AND HALVES

WRITING ASSIGNMENT

Write for an uninformed reader a 500-word detailed description of a mechanism, choosing a small mechanism from your technical area. The following are some suggestions:

carburetor	thyratron tube
piston	slide rule
IBM card	gyroscope
fireplace	diode
mechanical pencil	gate valve
fuel pump	a core module from a computer
thermostat	tachometer
barometer	opaque projector

Within your paper, you will inevitably make references to the mechanism's operation, particularly when you are stating the function of a part. However, make sure that your paper's emphasis is upon physical characteristics rather than operation. Describing a mechanism's operation is a separate technique of technical writing which will be explained in Chapter 4. To fully understand this distinction, compare the "Description of a Moving-Coil Meter" to the "Description of an Ammeter in Operation" at the end of Chapter 4. As you know, an ammeter is a type of moving-coil meter, so the two reports demonstrate the difference between describing physical characteristics and describing an operation.

EXERCISES

1. The frame of a dragster, like the frame of any car, is composed of side rails and crosspieces. However, the dragster differs from conventional cars in three main ways: (1) It is made of very light, somewhat springy material to transfer weight to the back wheels at the start of the quarter-mile journey. (2) Its rear axle, rather than being suspended from the frame, is welded to the frame. (3) The driver's compartment holds only one person and

has a roll cage to protect him. The frame, shown below, serves as a skeleton which holds every part of the completed dragster.

Write a detailed description of the dragster's frame. The rear axle housing has been excluded from the drawing to make the frame easier to describe. Aim your description at a reader who has never seen a dragster. Rely solely upon words to achieve communication: provide no visuals and assume that your reader does not have access to the drawing.

The following is some additional information about the dragster's frame:

> length: 17 feet
> width: front crosspiece—30 inches
> rear crosspiece—26 inches
> weight: 100 pounds
> material: seamless steel (aircraft) tubing
> method of construction: welding

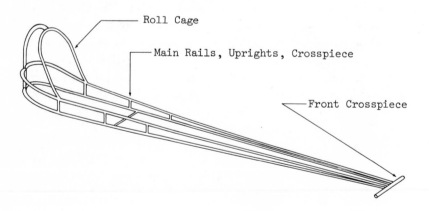

2. As shown on the following page, a cigarette lighter's three main parts are its outer casing, inner casing, and spring and screw assembly.

Outer Casing	Inner Casing	Spring and Screw Assembly
a. Case	a. Case	a. Spring
b. Cover	b. Wind baffle	b. Screw
c. Hinge	c. Striking wheel	c. Plug
	d. Latching device	

Write a detailed description of the outer casing and inner casing for an uninformed reader. Clarify the location and shape of the lighter's metal parts, but do not describe the wick, flint, or cotton filler. If possible, examine a lighter closely before beginning your description.

The following description of the spring and screw assembly will give you an idea of the amount of detail your description should contain:

```
The function of the spring and screw assembly is

to hold the flint against the striking wheel.  The

assembly consists of a spring, a screw, and a cylindri-

cal plug.  The spring is 1-1/4 inches long and 1/8 inch in

diameter.  Attached to one end of the spring is a plug

which is 3/16 inch long and 1/8 inch in diameter.  The

other end of the spring connects to a 1/4-inch-long metal

screw which has a slot across its top.  The entire as-

sembly fits plug first into a threaded tube within the

inner casing of the lighter.  When the screw is turned

into the end of the tube, it causes tension in the

spring, pushing the flint securely against the striking

wheel.
```

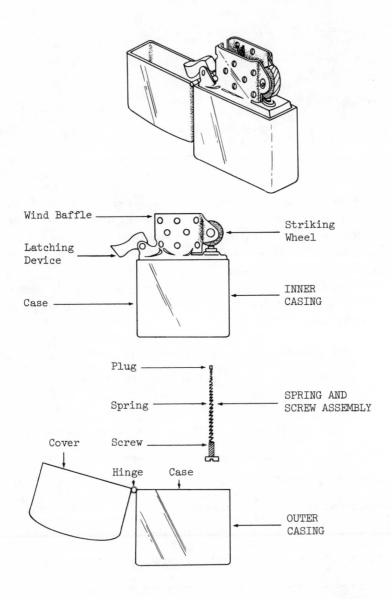

3. Write a detailed description of a ballpoint pen for an uninformed reader, emphasizing the shape and dimensions of each of the parts.

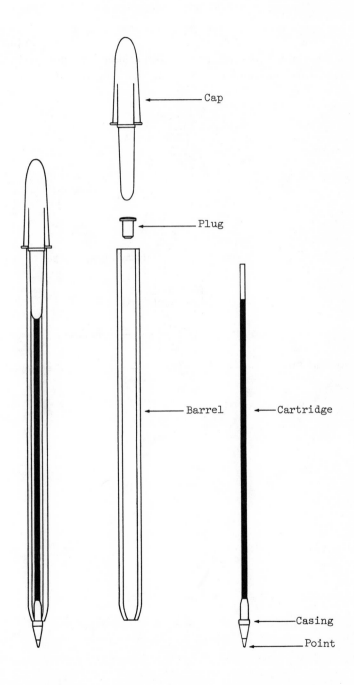

4. As concisely as possible, describe the exterior of a sparkplug for an uninformed reader. Precise dimensions are not necessary; concentrate on communicating the shape of each part.

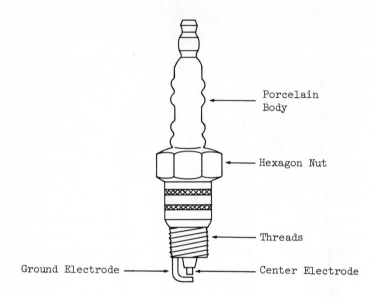

4. DESCRIBING A MECHANISM IN OPERATION

In our technical society, we depend upon the action of machines. Processes once accomplished by men with tools are now completed by machines under the control of men, and today's technical writing reflects that shift. *Describing a mechanism in operation* has become a more useful writing technique than the conventional *process description*, which emphasizes the role that men play. To put it another way, more and more buttons are being pushed all the time, and the action that occurs after the button is pushed takes precedence over the human activity of pushing the button.

Again in this chapter, the word *mechanism* includes all systems whose parts operate in a definable way. Thus, the principles for describing a mechanism's operation are applicable to almost all disciplines. Even nonhardware items such as athletic teams can be viewed as mechanisms. Successful teams are commonly referred to as "well-oiled machines" or "big red machines," and the comparison is valid. The power sweep, a play used by most football teams, involves eleven integral parts (players), each of whom functions with precision when the play is successful. Each player has an important job, if only to act as a decoy. The function of the power sweep is to pick up yards, to put six points on the scoreboard. And most teams subscribe to Vince Lombardi's "run to daylight" principle of operation, which means that the ball carrier heads upfield as soon as he gets an opening. Although this example is unusual, it should serve to emphasize the wide applicability of the principles for describing operations.

OUTLINE FOR
DESCRIBING A MECHANISM IN OPERATION

Description of an operation should not be confused with a physical description of the mechanism. The two are separate techniques of report writing and, as mentioned in Chapter 3, the physical description generally precedes a description of operation in industrial writing. The following outline is suitable for describing virtually any operation.

I. Introduction
  A. Definition of mechanism's function
  B. Explanation of operating principle(s)
  C. Statement of main sequences
II. Description
  A. Sequence one
    1. Definition of function
    2. The action
      a. Detailed description of action
      b. Results of action
      c. Relationship to other sequences
  B, C, etc.—same for remaining sequences
III. Conclusion (complete cycle of operation)

THE INTRODUCTION

In the introductory section, you must give the reader enough information to prepare him for the detailed description of operation. This requires defining the mechanism's function, explaining its principle(s) of operation, and identifying the sequences in the operation.

DEFINITION OF FUNCTION

It is perfectly permissible to tell the reader what the mechanism does by simply saying, "The function of a carburetor is to deliver the proper ratio of gasoline and air to the cylinders of an internal-combustion engine," or, "The function of a gate valve is to control the flow of liquid through a pipe." This can be followed by general information which might be helpful to the reader, such as the various places a gate valve is used.

EXPLANATION OF OPERATING PRINCIPLE

Basic principles of operation, such as the role gravity plays in a fuel-injection system or the function of vaporization in a carburetor, should be clarified in the introduction. By writing an operational definition of the principle, you give the reader all he needs and you avoid presenting equations and laws. The explanation of gravity, for example, would be limited to how it applies to fuel injection. If you were introducing the operation of a carburetor, your explanation of vaporization would resemble the following: "The carburetor distributes gasoline in the form of tiny droplets in an airstream. As a result of heat absorption on the way to the cylinders, these droplets are vaporized, making the mixture a flammable gas." This type of operational definition of the basic principle prepares the reader for a more complex description of the mechanism's operation in the body of the report.

STATEMENT OF MAIN SEQUENCES

Just as any mechanism can be divided into physical parts, any operation can be broken into sequences. For example, the four logical sequences of a piston's operation are its intake, compression, power, and exhaust strokes. In a description of operation, the mechanism must be analyzed in terms of its operation rather than according to its parts. For example, a description of the sequences of compression, condensation and evaporation is much more appropriate to the account of a refrigerator's operation than a description of the compressor, condenser, and evaporator units.

For most operations, the sequences need only be chronologically named in the introduction. When two sequences occur simultaneously, however, clarity demands that they be described separately in the body. Such information about the body of the report should be given in this section of the introduction.

DESCRIPTION

If the operation has been broken into logical sequences, each has a distinct action and function. The sum of the sequences described in the body should equal the mechanism's entire operation.

DEFINITION OF FUNCTION

You can generally describe a sequence in one paragraph, with the definition of the sequence's function serving as your topic sentence. To begin the description of a piston's intake-stroke sequence, for example, you would say that its function is to draw a fuel and air mixture from the carburetor through an intake-valve opening and into the piston cylinder. You would then complete the paragraph with a detailed description of the intake-stroke action.

THE ACTION

As shown in the outline, this section requires description of each action in the sequence, the results of each action, and the relationship of each main sequence to others. In the chain reaction of any mechanism's operation, the result of one action is the cause of another. Therefore, a detailed description amounts to a series of cause-and-effect statements. These elements are identified in the following description of a piston's intake-stroke sequence:

During the intake stroke, the piston moves from the top of the cylinder to the bottom [ACTION]. This downward motion creates a partial vacuum [the vacuum is a RESULT of the piston's ACTION] which draws [ACTION of the vacuum] the fuel-air mixture around the open intake valve and into the cylinder [RESULT of the vacuum's ACTION]. As the piston reaches the bottom, the inlet valve closes [ACTION], locking a fresh charge in the cylinder [RESULT of the valve's ACTION], and completing the first cycle of the four-cycle engine [RELATIONSHIP TO OTHER SEQUENCES].

At the end of the intake-stroke sequence, a valve has locked the fuel in the cylinder. To tell the reader what happens next, the compression-stroke sequence, start a new paragraph with a topic sentence: "The compression stroke compresses the fuel mixture against the top of the cylinder, making the fuel volatile." Then, using the technique shown above, provide a detailed description of the sequence. This writing process continues until each sequence in the operation has been described. The operation of any mechanism is extremely logical, and so is the technique for describing it. To successfully explain an operation, you must observe the mechanism carefully and record what happens as clearly as possible.

THE CONCLUSION

In the conclusion, take the reader through a brief, one-paragraph description of the mechanism's complete cycle. Details are not necessary because the operation has already been divided into sequences and thoroughly described; it must now be put back together, or synthesized, for the reader. You can accomplish this by tying the topic sentences of each sequence into a general statement of the mechanism's entire operation.

A piston's operation consists of four strokes. During the intake stroke, the descending piston draws a fuel-air mixture into the piston cylinder. As the piston rises, its compression stroke forces the mixture against the top of the cylinder and makes it volatile. A spark from the sparkplug ignites the mixture, and combustion drives the piston back down the cylinder, turning the crankshaft, during the power stroke. As the piston ascends again, its exhaust stroke forces the spent gases from the cylinder.

ILLUSTRATING THE ACTION

When describing an operation, you achieve your objective if you enable the reader to visualize the action. Effective use of illustrations simplifies this task by reinforcing the paragraphs of description. As Chapter 7 emphasizes, you have some options in selecting the type of illustration which will best serve your reader. Simple block or flow diagrams are useful for symbolizing overall sequences of action, and a drawing like the one on the following page can help clarify the action within each sequence. A good visual gives the reader no more and no less than he needs. For most readers, drawings and simple diagrams are much more meaningful than highly symbolic schematics.

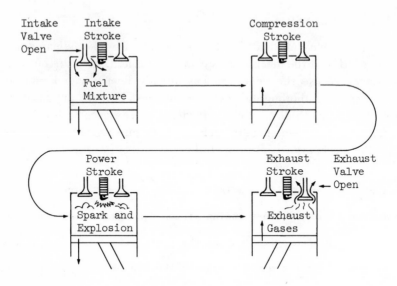

ADVANTAGE OF THE ACTIVE VOICE

Logic demands that you use the active voice when describing the action of a mechanism. The following is an abbreviated discussion of voice; the "Voice" section of the Appendix goes into greater detail about intransitive and transitive verbs. "Voice" refers to the relationship between the subject and verb of a sentence. In an active sentence, the subject does something; a passive sentence, on the other hand, allows the verb to act upon the subject.

Passive: The flint is struck by a turning wheel.
Active: A turning wheel strikes the flint.
Passive: The wheel is turned by force.
Active: Force turns the wheel.
Passive: An explosion in the cylinder will be caused by the compression of gases.
Active: The compression of gases will cause an explosion in the cylinder.
Passive: The current is measured by the reaction of a pointer to a permanent magnet.
Active: A pointer, reacting to a permanent magnet, measures the current.

All the parts are acting during the operation of a mechanism, and the active voice pinpoints their action, making the operation easier for the reader to grasp. The passive voice wastes words and often confuses the meaning of sentences. Whenever possible, use the active voice, not only for describing operations but for any type of technical writing.

When reading the three student models which follow, notice how they concentrate on each mechanism's operation and deemphasize its physical characteristics. In the first paper, for example, the writer's first heading is "Action of the Permanent Magnet" rather than "Permanent Magnet," and the paragraph beneath that heading emphasizes operation rather than physical description. In the second report, the first heading is "Conversion to Electrical Energy," and the paragraphs stress this operation while providing just enough physical description to make the operation understandable. Although the third model involves an operator (a pilot), it emphasizes the forces which affect the movements of an airplane.

Each of the models has its strong points, but keep in mind that none of them is perfect. Neither is any other model in this text. Examine each of them critically, and offer your classmates specific, positive suggestions for improving them.

DESCRIPTION OF AN AMMETER IN OPERATION

INTRODUCTION

The function of an ammeter is to measure the amount of current flowing in an electrical circuit. In order to operate, the ammeter must be placed in a circuit so that current can flow through it. The combination of a magnet created by the flow of current and a magnet already in the ammeter causes the meter needle to move. This movement indicates the presence of current and also measures the amount of that current.

The action of the permanent magnet will be discussed, followed by a description of how a magnetic coil is created. Then, the combined operation of the two magnets and the resulting measurement of current will be described.

ACTION OF THE PERMANENT MAGNET

A magnet has magnetic poles, one north and one south, at its respective ends. If a pole happens to be north, it attracts south poles and repels other north poles. A south pole attracts north poles and repels other south poles. In

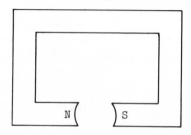

FIG. 1   PERMANENT MAGNET

an ammeter, the magnet which has these characteristics is called a permanent magnet (Fig. 1). It is one of two magnets needed for the operation of the ammeter.

CREATION OF THE MAGNETIC COIL

Electric current flowing through a coil of wire can create a magnet. This happens with the moving coil in an ammeter (Fig. 2). As current flows through the coil, a north pole forms at one end of the coil and a south pole

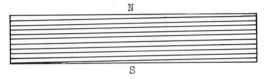

FIG. 2   MOVING COIL

forms at the other end. These magnetic poles work the same way as magnetic poles on a magnet: like poles repel each other (north will repel north) and unlike poles attract each other (north will attract south). This magnetic coil is placed within an opening in the permanent magnet (Fig. 3), and a needle is attached to it.

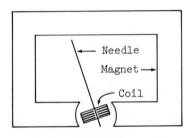

FIG. 3   MAGNET AND COIL

OPERATION OF THE AMMETER

When the current is connected, current flows through the ammeter. While doing so, it flows through the moving coil and creates the north and south poles. The magnetic poles at the ends of the permanent magnet have already been formed, so two north poles and two south poles are now near each other, as shown in Figure 4. The like poles repel each other.

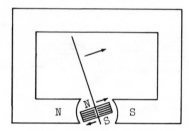

FIG. 4   AMMETER OPERATION

The amount that the poles repel each other depends on the strength of the poles at each end of the coil. The strength of the poles, however, depends on the amount of current that flows through the ammeter. As the current becomes larger the strength of the poles increases and the amount that the poles repel each other also increases. The repulsion causes the moving coil and the indicator needle to deflect. The needle deflects farther as the current increases. When an operator disconnects the ammeter from the circuit, the current stops flowing, the coil becomes demagnetized, and the needle returns to zero.

-3-

CONCLUSION

Basically, an ammeter contains a permanent magnet and a coil. The coil, which has a needle attached to it, becomes magnetized when an operator connects the ammeter to a source of current. These two magnets repel each other. The repulsion deflects the needle, which indicates the amount of current flowing. When an operator disconnects the ammeter from its source of current, the coil demagnetizes and the needle returns to zero.

OPERATION OF
THE BASIC HIGH-FIDELITY MUSIC SYSTEM

INTRODUCTION

The function of a high-fidelity system is to electrically reproduce prerecorded sounds. Three sound sources available are radio broadcasts, phonograph records, and magnetic-tape recordings. This paper is limited to the operation of a high-fidelity phonograph-record system.

The three main sequences in the system's operation are conversion of the record to electrical energy, amplification of the electrical signals (dealt with briefly in this report), and conversion of the electrical energy to mechanical energy needed to produce sound waves.

CONVERSION TO ELECTRICAL ENERGY

A record is simply a physical representation of an electrical signal, which itself is a representation of a sound. This physical representation exists as a series of ripples in the sides of a V-shaped groove cut into the record disc. These ripples cause the stylus (record needle) to vibrate as it slides along the groove of a rotating record; in other words, the stylus feels the physical representations of various sounds. It transmits the vibrations to the phonograph's magnetic cartridge.

Basically, a magnetic cartridge is a miniature electric generator that converts mechanical energy into electrical

energy. It has a tiny permanent magnet and a miniature wire
coil (Fig. 1). The coil, when moved through the magnet's
lines of force (between its poles), generates electrical
current. The current is proportional to the speed of the
coil; faster movement of the coil produces more current.

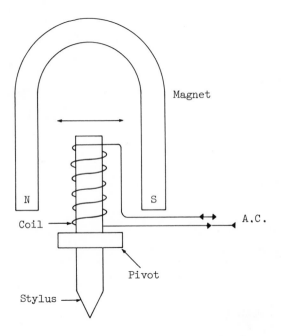

FIG. 1   MAGNETIC CARTRIDGE

In a phonograph's moving-coil cartridge, the coil attaches to the stylus. As variations in the record groove
cause vibrations in the stylus, the coil moves through the
magnet's lines of force at different speeds, creating
various electrical signals. The signals produced by this

method are very weak (0.01 to 0.2 volt) and must be amplified before reaching the phonograph's speaker system.

AMPLIFICATION

The amplifier takes the weak signals from the phono-
cartridge and gives them the power needed to drive a speaker
system. The amplifier also boosts the bass and treble notes.
The low bass notes and very high treble notes need boosting
because the human ear is not as sensitive to these sounds as
it is to sounds of the middle-frequency range. When adjusting
the volume, the operator controls the strength of signals
sent to the speakers.

SOUND REPRODUCTION

After the electrical impulses have been strengthened in
the amplifier, they are directed to the speakers. The speakers are designed to convert electrical energy into the mechanical energy needed to produce sound waves. A speaker contains
a huge permanent magnet, a wire-wound coil known as the voice
coil, and a treated paper or plastic cone attached to the voice
coil (Fig. 2). Electric impulses create a magnetic field as
they pass through the voice coil. This field constantly
changes direction as the signal changes direction (alternating
current). This change in direction of the coil's magnetic
field causes it to be attracted to and also repelled from the
permanent magnet mounted behind it. This piston-type action
of the voice coil also moves the cone back and forth. The

-3-

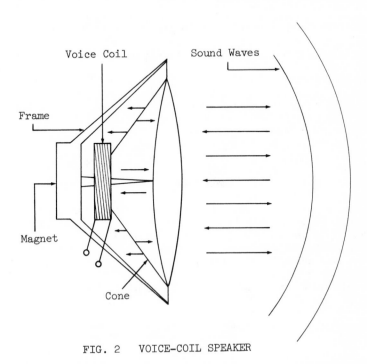

FIG. 2   VOICE-COIL SPEAKER

cone acts as a pump, pushing out air waves which are heard
as sound.

Hi-fi speakers must be able to reproduce the sound frequencies that can be detected by the human ear. These frequencies range from about 20 cps (cycles per second) to
20,000 cps. The cone of a bass speaker vibrates relatively
slowly, moving large amounts of air to produce low bass
tones. To generate high-frequency treble tones, the speaker
cone must race back and forth at a very rapid rate (up to
20,000 times per second). Because speakers must produce
high and low notes simultaneously, the work load is usually

divided between two or three special speakers designed to handle either high, medium, or low-frequency sounds. Special speakers for bass sounds are called woofers; those designed for high notes are called tweeters; the mid-range sounds are handled by mid-range speakers.

The signals are sent to the speakers over the same two wires, highs and lows mixed together. The various frequencies must be sorted out and directed to their intended speakers. This is done by an electronic circuit known as a crossover network. Located in the speaker enclosure, this circuit acts as an audio traffic cop, directing the different frequencies to the proper speaker.

CONCLUSION

The phonograph's magnetic cartridge converts the mechanical movement of the stylus into electrical energy. An amplifier strengthens the cartridge's impulses and sends them to the speaker system. The speaker system converts the electrical impulses back to mechanical sound energy, releasing high-fidelity sounds over a wide range of the sound spectrum.

THE BASIC MOVEMENTS OF AN AIRPLANE

INTRODUCTION

This report explains how a pilot controls the three basic movements of an airplane. For the purpose of this report, a single-engine, propeller-driven monoplane (single wing) will be used.

The pilot controls a plane with the wheel and rudder pedals. Through a system of cables and pulleys, these controls connect to the plane's control surfaces, which include the rudder, ailerons, and elevators shown in the drawing below. Rudder pedals on the floor of the cockpit control the plane's rudder. The wheel mounted on a column protruding from the instrument panel controls the ailerons and elevators; moving the wheel forward or backward raises or lowers the elevators; turning the wheel right or left controls the ailerons.

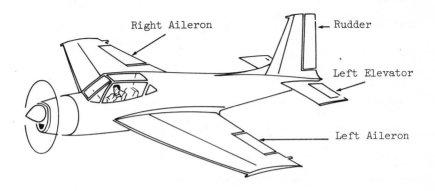

FIG. 1   CONTROL SURFACES

Through the pilot's adjustment of these controls, and
the resulting movement of the control surfaces, a plane performs three basic movements: yaw, roll, and pitch.

YAW

Yaw is the swinging motion of the plane on its vertical
axis. This happens when the pilot presses the right or left
rudder pedal, depending on the direction of yaw desired.
Pressing the right rudder pedal moves the rudder to the
right, and the force of air against the rudder swings the
nose of the plane to the right. The rudder functions in the
manner of a boat's rudder.

ROLL

Roll is the motion of the plane on its longitudinal
axis. Pilots call it "banking." The pilot turns the wheel
in the direction he wants the plane to roll (right, for
example). This action causes the left aileron to move down
and the right aileron to move up. The air pushing against
the ailerons forces the right wing down and the left wing
up. The degree the pilot turns the wheel controls the
degree of roll. To return the plane to level flight, he
turns the wheel in the opposite direction.

To change the direction of the plane, the yaw and roll
movements are combined. When making a right turn, the pilot
simultaneously pushes the right rudder pedal and turns the
wheel to the right.

PITCH

Pitch, the nosing-up-or-down motion on the transverse axis, is caused by pushing the wheel forward or pulling it backward. Pushing forward moves the elevators up, which causes the nose of the plane to go up. Pulling the wheel back makes the elevators go down; the air pushes against the elevators, forcing the tail up and the nose down. The pilot uses this movement for ascending or descending.

CONCLUSION

A short trip will clarify how these operations fit together. The spinning propeller creates more and more thrust (the force, caused by the propeller's "bite," which pulls the plane), moving the plane forward on the runway. When the plane attains flying speed, lift occurs. (Because of the wings' shape, air flowing over them must move faster and farther than air flowing under them, creating greater upward force.) After the plane leaves the ground, the pilot pulls the wheel back slightly, and the plane begins to climb.

As the plane reaches the desired altitude, the pilot pushes the wheel forward, leveling the plane, and returns the wheel to neutral. Then, he simultaneously turns the wheel to the right and pushes the right pedal, thereby tilting the wings to the right and swinging the nose of the plane toward the right. Just before reaching the desired direction, he applies the opposite controls, straightening the plane, and returns the controls to neutral.

When the plane approaches the airport, the pilot pushes the wheel forward, dropping the nose, and the plane begins its descent. A few feet above the ground, he reduces the speed. The plane loses its lift and settles to the ground.

WRITING ASSIGNMENT

Write a 500-word description of a mechanism in operation for an uninformed reader. Select a mechanism from your technical area, and avoid operations which require extensive participation by an operator. Your objective is to *describe* an operation rather than *direct* one; if you find yourself writing a set of directions, choose a better topic. Some possible topics are listed below:

carburetor	air conditioner
diode	electric soldering gun
IBM sorter	disc brakes
circuit	electric fan
fuel-injection system	ignition system
arch	thermostat
gate valve	electric gasoline gauge
compass	tachometer
speedometer	capacitor

EXERCISES

1. First, write a one-paragraph description of a paper clip, clarifying its shape and dimensions. Then devote as many paragraphs as necessary to a detailed description of how a paper clip operates. Your paper will have to explain torsion, the principle by which the clip works. Aim your writing at a technically uninformed reader.

2. Immediately after jumping from an airplane, a parachutist free falls at over 100 miles per hour. When he pulls the ripcord and the canopy rapidly fills with air, he experiences the opening shock and begins descending steadily at approximately 15 miles per hour. In technical terms, he enters a state of equilibrium because the resultant force of drag and lift equals the weight.

   Explain this state of equilibrium to a reader who is familiar with neither the forces involved nor technical terminology. Consider using a diagram to reinforce your explanation.

3. Write a detailed description of what happens when the driving gear turns, including an explanation of the gear ratios. The ratio between the driving gear and the idler gear is 1:2, and the ratio between the idler gear and the pinion gear is 2:3. Aim your description at a technically uninformed person.

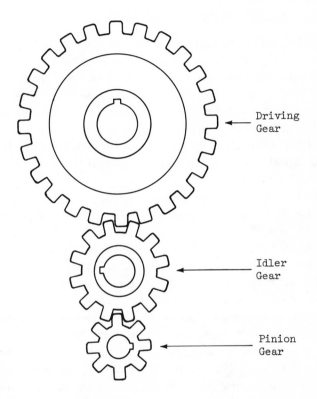

4. Write a detailed description of the action that occurs in the backfield during the football play diagrammed below. Clearly explain the quarterback's options, dividing the play into sequences if necessary. Aim your description at someone who knows nothing about football.

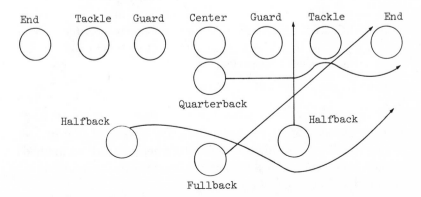

5. INTERPRETING STATISTICS

Interpretation of statistical data is the process of deriving and communicating logical conclusions from a body of facts. A widely held misconception is that data, or facts, somehow speak for themselves. If they do, they speak with ambiguity because two people examining the same data often arrive at totally different conclusions. Until they are interpreted by the report writer, statistical data are nothing but ink marks on pieces of paper. Through interpretation, the facts do speak, and if the writer has been rigorously logical, they speak with authority.

Although most of us are not fully aware of it, we interpret data every day. When deciding whether to pass a truck, we consider the acceleration of our car, the speed of the truck and oncoming cars, the road, the visibility, and perhaps how late we are for our meeting, before making a decision. Many of our everyday choices, some more crucial than others, require such interpretation, and so do the daily decisions in business and industry. Those in management and supervisory positions are constantly making moves which affect the profit and ultimately the survival of their firms. The problems range from critical matters such as whether or not to expand, to develop a new product, or to install a new system, all the way down to such routine situations as promoting the most qualified man, deciding whether a welded or soldered joint is better for a certain structure, or selecting the best brand of fork lift.

Interpreting statistics is one of four steps you take as you prepare a recommendation report. Although the steps overlap each other, the recommendation report-writing process always involves collection, analysis, presentation, and finally, interpretation of data. Depending on the complexity of the data, this process can require simply asking a few questions and writing a short recommendation report, or spending months sifting through data and compiling a formal feasibility report.

1. COLLECTION OF DATA

Generally speaking, the more information examined, the greater the probability of making the right recommendation. Four methods of data collection are observation, testing, interviewing, and studying printed material.

Direct observation is probably the most common way of getting information. For example, to figure out when a shipment will be ready for a customer, a writer observes and measures the time it takes to manufacture a certain number of parts; he then uses that data to estimate the completion time for the entire order. As in all methods of information gathering, the observer has to insure that his data are accurate. He must consider the possibility of equipment breakdowns, and estimate their potential effect on the production rate.

Testing is similar to observation, except that the conditions are more tightly controlled. For example, the strength of various types of steel can be tested to determine the best type for a particular bridge. The testing occurs in a laboratory, where technicians create conditions identical to those the steel must withstand when the bridge is in use.

Conducting interviews is another effective method of gaining information. Anything from carrying on an informal discussion with a fellow employee to interviewing a local authority may prove helpful. A writer who must recommend the best brand of equipment can request opinions from other firms. Interviews and surveys do not always yield statistical data, but they generally provide evidence to support the writer's recommendation.

Printed material serves as an excellent source of data for recommendation reports. Technical journals contain information about contemporary trends and innovations, precisely the type of data a writer needs to make recommendations affecting the future of his department or firm. Sales literature from manufacturers is also available, and many firms have libraries on technical subjects. Chapter 6 provides detailed information about collecting and using printed material.

2. ANALYSIS OF DATA

The only effective way to make data meaningful is to analyze them, or separate them, according to criteria. Selection of logical criteria is crucial to the entire recommendation report, because the criteria are the bases on which the recommendation will ultimately be made. To explain how they are selected, we can create a hypothetical situation.

Let's say that a steel mill is having problems with its radio units. The present tube-type radios, which are supposed to provide voice communication between workers on the floor and crane operators high overhead, are breaking down too often, thus causing high repair bills and lost production time. The breakdowns are occurring because the tube radios cannot withstand the wide temperature variations within the mill and the occasional high surges of voltage, called "spikes," which are common in steel mills.

Six months previously, a new transistor radio appeared on the market. Sales representatives claimed that it could survive in a steel-mill environment, so two units were installed on a trial basis. The trial period has now ended and you must write a report recommending that either tube or transistor units be purchased in the future. In order to reach this decision, you decide to compare the performance of the two transistor units to two tube units which have been in service for approximately six months. You find the following types of data for the four units: initial cost, number of maintenance calls, number of man-hours for maintenance, cost of components replaced, cost of labor for repairs, and production hours lost during breakdowns.

The six items are tentative criteria by which to judge the two systems. During examination of the data, you discover that there is very little difference in the "initial cost" of the alternative units, so you reject that criterion. Further study indicates that the "number of maintenance calls" and "number of man-hours for maintenance" are insignificant because they are covered by the "cost of labor for repairs" criterion. The criteria for judging the systems, or more specifically, for judging the data collected for each of the systems, are finally narrowed to (1) cost of labor for repairs, (2) cost of components replaced, and (3) production hours lost during breakdowns. You must now separate the data, or analyze them, according to the criteria selected.

Criteria must always be chosen carefully because of their tremendous influence on the recommendation. Suppose, in the above example, you recommended the transistor radios because their components were half as expensive, but later learned that the transistors took three times the production hours to repair. The recommendation would have been a costly one because you failed to consider the criterion of production hours lost during repair.

3: PRESENTATION OF DATA

In writing the report, you must decide whether to present the statistical data in paragraphs or in visuals. The example below shows what data for the hypothetical radio problem would look like if presented in paragraph form:

Maintenance costs for one of the transistor units totaled $392, of which $305 was for labor and $87 for parts. The cost for repairs to the second transistor unit was $272, with $89 of that going for parts. The first tube radio cost $256 to repair, of which $33 was for parts. The second unit cost $143 for labor and $17 for parts, a total of $160.

Twenty-six production hours were lost because of breakdowns of the first transistor unit and 20 were lost with the second transistor unit, while tube units one and two lost 13 and 9 production hours respectively.

The same data are presented in the table below:

TABLE 1   MAINTENANCE COST AND PRODUCTION HOURS LOST

Item	Trans. 1	Trans. 2	Tube 1	Tube 2
Cost of Parts ($)	87	89	33	17
Cost of Labor ($)	305	183	223	143
TOTAL ($)	392	272	256	160
Production Hours Lost	26	20	13	9

There is little question of the visual's superiority for presenting statistical data. Readers have difficulty grasping statistical comparisons within paragraphs, but find that a good table makes the data readable and understandable. The effort required to convert data into effective visual form pays off in communication.

4. INTERPRETATION OF DATA

After presenting the data, you must interpret them, or tell the reader what they mean. Interpretation of data amounts to pointing out important relationships that are suggested in the data and emphasizing particularly significant bits of data. Good interpretation does not repeat what has already been presented; rather, it integrates the data and your ideas about the data. As suggested earlier in this chapter, it is dangerous to assume that the reader will get everything out of the data that you intend, particularly if the data are complex. Although the data from the radio problem are fairly uncomplicated, the following paragraphs show how the information presented in the visual might be interpreted:

Maintenance Costs

In order to compare maintenance costs of the transistor and tube systems, two units of each type were installed and tested for 6 months. The test results (Table 1) show that the present tube radios cost less to maintain. The transistor units averaged $322 for parts and labor, versus $208 for the tube radios. Despite sales claims, the data indicate that transistor units are unable to withstand the mill environment. Failure reports show that the transistors are more sensitive to temperature and voltage extremes.

Production Hours Lost

Crane operators who have used the transistors say that while the units are working, they perform better than the tube-type radios; however, the transistors simply do not work enough of the time, as indicated in Table 1. During the 6-month period, 46 production hours were lost during repair of the transistors, versus 22 hours lost because of breakdowns in the tube radios.

As emphasized in Chapter 7, "Illustrating," visuals must always be directly referred to in the interpretation. Also, a visual should not appear before its interpretation. Refer the reader to the visual near the beginning of the interpretation; in doing so, you insure that he will read the interpretation, and you control his path through the data. You can do this effectively if the visual appears after the interpretation or between paragraphs of interpretation, but not if the visual precedes its interpretation.

OUTLINE FOR INTERPRETING STATISTICS

The principles for interpretation of data are applicable to both informal recommendation reports and formal feasibility reports. Feasibility reports, discussed in Chapter 11, are reserved for problems requiring interpretation of extremely complex data. Recommendation reports, which are more common and much shorter, can be organized according to the following outline:

I. Introduction
   A. Purpose of report
   B. Definition of problem
   C. Scope
      1. Alternatives
      2. Criteria

II. Presentation and Interpretation of Data
   A. Judgment according to first criterion
      1. Explanation of criterion
      2. Presentation of data for all alternatives
      3. Interpretation of data

    B. Judgment according to second criterion
       1. Explanation of criterion
       2. Presentation of data for all alternatives
       3. Interpretation of data
    C, D, etc.—same as above for rest of criteria
III. Conclusions
IV. Recommendation

THE INTRODUCTION

Although the introduction should be as concise as possible, you cannot risk sending the reader into the body of the report unprepared. Therefore, you explain the report's purpose, define the problem at hand, and state the scope of the report.

PURPOSE OF REPORT

There is absolutely nothing wrong with beginning the report by saying, "The purpose of this report is. . . ." The purpose, of course, is to recommend a solution to a particular problem, and you can generally state the purpose in one sentence.

DEFINITION OF PROBLEM

The statement of purpose leads you directly into a definition of the problem, which is essentially an explanation of what is wrong or what has made a report necessary. In the hypothetical situation that has been used throughout this chapter, the problem is that the present radio system is breaking down too often and costing too much in terms of repair bills and lost production time. This, of course, is a simplified explanation; the reader needs background information, particularly concerning the cause of the breakdowns, to clearly understand the problem. If he does not understand the problem, he may not clearly understand the information in the body of the report, either.

SCOPE

After defining the problem, your next logical step is to state the alternative solutions and the criteria used. Occasionally, the alternatives can simply be named, but generally some elaboration is needed. In the radio problem, a brief description of the two radio systems, emphasizing their differences, would help the reader understand the comparisons made later in the report.

The final item in the introduction is a statement of the criteria. To indicate how you have organized the data, name the criteria in the order they will appear in the body of the report. The criteria often require explanation, but reserve this for later in the report when you deal with them individually.

PRESENTATION AND INTERPRETATION OF DATA

This section of the report was discussed and exemplified earlier in the chapter. It should be emphasized, however, that each criterion is named in a main heading in the body of the report. Immediately after each main heading, provide an explanation of the criterion if one is necessary. For example, if data in the cost section were gathered during a six-month period, you would give the reader that information. You would also specify the types of data included in the cost section; the criterion could refer only to initial costs, or could cover costs for installation, operation, and maintenance.

A "Conclusions" section appears at the end of the report, so your comments in the body of the paper should be interpretive rather than conclusive. Interpreting data according to a criterion naturally implies certain tentative conclusions, but you do not want to overemphasize conclusions before all of the evidence has been presented.

THE CONCLUSIONS

After presenting and interpreting data for all the criteria, you are ready to draw conclusions. Rather than restating what you have previously said about the individual criteria, your conclusions should be comprehensive. The reader wants to know what the data mean in terms of the overall problem. Therefore, this section should contain explicit statements which emphasize particularly significant points and which prepare the reader for the final recommendation.

THE RECOMMENDATION

If your recommendation surprises the reader, your report fails. Carefully written reports build steadily toward the recommendation, often reducing it to a formality. For short reports, one or two sentences should suffice, but for complex reports which involve many aspects of a problem, a paragraph may be necessary.

DANGERS WITH STATISTICAL DATA

1. Be careful of bias. After making sure the sources of the data are good, let the data go to work for you in your effort to make the right decision. If you happen to make the wrong recommendation based on careful interpretation of data, you at least have the data to back you up. If you base a wrong recommendation on your own preconceptions, you have nothing.
2. The data which support your recommendation must be valid. The fact that a particular system works fine for X Company does not necessarily mean the system will work for your firm. Perhaps X Company is larger, has a lower calibre of employees, and emphasizes quantity rather than quality. These things make a difference.
3. Do not assume that because alternative A is bad, alternative B must be good. Maybe both are bad. Also, do not assume that A should be recommended simply because A is good. Maybe B is better. Examine each of the alternatives thoroughly and objectively.

The following student models exemplify methods of approaching various problems. In the first, the writer evaluates alternative mechanisms and recommends that one be purchased. Notice how the report is organized. In the second report, which concerns a pollution problem, the statistical data are not presented in tables. Decide whether the data are complex enough to demand visual presentation.

RECOMMENDATION OF A MAGNETIC STORAGE DEVICE

INTRODUCTION

PURPOSE

The purpose of this report is to recommend a magnetic storage device for the Harris Department Store.

PROBLEM

The store is considering the possibility of installing a data-processing system, and needs a storage device which will provide information on accounts-receivable files as quickly as possible.

SCOPE

Two alternative storage devices, magnetic tape and magnetic disk, are evaluated in this report. The most important aspect of the storage device is the time it requires to obtain data (access time). Because of the store's size, the capacity of the system is of secondary importance.

ACCESS TIME

A magnetic-tape unit has sequential-access files, which means that each record must be read in sequence. For example, if record 05 is to be read, records 01 through 04 must be read first, slowing the access time. The time varies according to the number of records which must be read before reaching the desired record.

The magnetic-disk unit uses random-access files. Random access refers to the ability of the system to skip around within the file and read or write specific data with no regard to sequence. This allows much faster access to data stored on magnetic disks than data on magnetic tape.

CAPACITY

The capacity of a storage device is expressed as the decimal digits, alphabetic characters, and special characters it can store at one time. Magnetic tape, which works much like a home tape recorder, will accept magnetic recordings. The tape consists of five to eight channels in a row, with each row containing one character of information. Reels of tape vary in capacity from 100 to 800 characters per inch, which amounts to 1.9 million to 3.6 million characters of information if an entire tape is filled with data.

A magnetic-disk unit stores characters on a disk resembling a phonograph record. The unit looks like a juke box, with a number of stacked disks and a movable read-write head capable of contacting either side of the disk. Each disk holds 200,000 characters of information, and a magnetic-disk storage device usually consists of 50 disks which can store 10 million characters.

At first glance, the table below seems to indicate that the magnetic-disk unit has more storage capacity than the

-2-

magnetic-tape unit. However, it is possible to change reels on the magnetic-tape unit, giving it unlimited storage capacity.

CAPACITY OF STORAGE UNITS

Item	Magnetic Tape	Magnetic Disk
Density	100-800 Characters per inch	100,000 Characters per disk face
Maximum Characters Per Unit	3.6 million	200,000
Maximum Storage	Depends on number of reels	Depends on number of disks

CONCLUSIONS

Magnetic tape provides the greatest storage capacity, using a number of reels of tape. However, access time is of prime importance, and the magnetic disk provides immediate access to data.

RECOMMENDATION

Because of its faster access time, the magnetic-disk storage unit should be installed in the Harris Department Store data-processing system.

RECOMMENDATION OF AN INCINERATOR SYSTEM

INTRODUCTION

PURPOSE

The purpose of this report is to recommend an incinerator system for the city of Rockton.

PROBLEM

Rockton has always used landfilling to dispose of solid wastes such as garbage, other trash from homes, and waste from commercial industries. However, the city is running out of space for waste disposal and must now select an effective method of incineration.

SCOPE

This report compares conventional incineration to the recently developed high-temperature incineration. These methods are evaluated according to efficiency, cost, and air pollution.

EFFICIENCY

The conventional incinerator burns the waste at a temperature of 2,000 degrees Fahrenheit, which reduces the volume of waste by 70 percent. These incinerators are capable of reducing 1,000 tons of waste per day. Most of the reduced waste must still be disposed of by landfilling, but some of it can be converted into fertilizer.

The high-temperature incinerator burns waste at a temperature of 3,000 degrees Fahrenheit. All waste, including solid materials and metals, is either burned or melted. The unit reduces volume by 97 percent. Remaining waste is disposed of in the same manner as conventional-incinerator waste.

COST

Conventional incineration is the most common type of incineration at the present time. The initial cost of a conventional incinerator is $3 million. The cost of operating these incinerators is $7 to $9 for each ton of waste. This amounts to a cost of $7,000 to $9,000 per day.

The high-temperature incinerator is a relatively new development. The initial cost of these incinerators is $1.5 million. The cost per ton for operating the device is $6 to $8. This results in a daily expenditure of $6000 to $8000.

AIR POLLUTION

The main drawback of the conventional incinerators is air pollution. Small particles and gaseous emissions are released from the incinerator during operation. Air pollution devices, however, have reduced the pollution by 97 percent.

The high-temperature incinerators require less oxygen for burning, which reduces its air pollution considerably.

With the installation of air pollution devices, the amount of pollution is almost zero.

CONCLUSIONS

High-temperature incinerators cost $1.5 million less than conventional incinerators and cost $1000 less per day to operate. In addition to doing a slightly better job of reducing air pollution, high-temperature units burn the waste 27 percent more effectively. A majority of cities in the United States use the conventional system, but only because they purchased the system before high-temperature incinerators were developed.

RECOMMENDATION

Installation of a high-temperature incinerator system by the city of Rockton is recommended.

WRITING ASSIGNMENT

Assume that you are working for a local firm and have been asked to evaluate two alternative types or brands of equipment and to recommend that one be purchased. In selecting alternatives, do not "stack the deck" too much in favor of one or the other. Each of the alternatives should be workable; your problem is to recommend the one which will work better. After deciding on criteria by which to judge the alternative pieces of equipment, collect the necessary data and write a 500-word recommendation report for an uninformed reader.

EXERCISES

1. Statistical data often become needlessly confusing when they are presented within a paragraph. Sort out the statistics in the following paragraph and place them in their more natural environment, a table.

    The turntable manufactured by ABC Company costs $119.95, versus $135.50 for a comparable XYZ unit. The base (cabinet) for ABC's turntable is priced at $7.50, and the cartridge sells for $39.95. XYZ Company's cartridge costs $37.50, and the cabinet is priced at $10.00. The companies charge the same amount, $7.50, for plastic dust covers. The entire ABC turntable assembly costs a total of $174.90, while an XYZ unit costs a total of $190.50.

2. Who won? Statistics are notorious liars, but careful analysis and interpretation of the following data from a college football game should enable you to reach a logical conclusion. Start by narrowing the number of criteria by which to base your decision. For example, the "yards penalized" criterion can be rejected because the teams' statistics are not significantly different. Do not

be confused by the passing statistics: team X attempted 49 passes, completed 29, and had 2 intercepted. In analyzing the data, you may decide to create a main criterion of your own and to make sub-criteria of some of the existing criteria.

Write your interpretation in paragraph form, and define potentially confusing football terminology. During class discussion after completion of this exercise, defend your decision.

	Team X	Team Y
First Downs	26	26
Yards Rushing	70	396
Yards Passing	308	74
Return Yardage	3	37
Passes Attempted	49	19
Completed	29	9
Intercepted By	1	3
Punts	6	4
Average	35	42
Fumbles Lost	1	0
Yards Penalized	72	70

3. Assume that you have decided to buy a motorcycle and have narrowed your selection to comparable models of Yamaha and Honda. First, determine criteria by which to judge the two cycles. Some possible criteria include (1) cost, (2) horsepower, (3) number of gears, (4) performance, (5) durability, (6) and special features. The criteria you select will depend upon what you are looking for in a cycle, and may differ from those chosen by others. After selecting criteria, gather pertinent data about the two cycles. Information can be gained from sales brochures, dealers,

and magazines. Finally, write a report recommending the better cycle. Organize the report according to criteria. If it contains complex, potentially confusing statistics, present them in tables. Do not let preconceptions or biases get in the way of accurate interpretation of the statistics.

4. Assume that a friend of yours, someone interested in pursuing a bachelor's degree in your technical field, has asked you for advice. He is trying to decide whether to enroll at a four-year school several hundred miles from his home, or to attend a community college for two years before transferring to the four-year school.

To assist him, evaluate the alternatives (using real schools) and write a recommendation report. Your first task is to select logical criteria upon which to base the recommendation. The criteria may include the total cost of attending each school, the availability of part-time jobs, the transfer of credits to the four-year school, or any others that you think are important.

Gather data for each of the alternatives and structure the report according to criteria. All the data need not be statistical but each of your ideas must be supported by facts. To insure that your friend clearly understands the information, consider presenting complex data in visual form.

6. RESEARCHING PUBLISHED INFORMATION

For technical students, the technique of researching published information has value beyond the writing of acceptable term papers. Technologies produce volumes of new material daily, making knowledge of the library useful for experienced professionals as well as students. Published information helps keep industry informed about up-to-date methods, equipment, and products. That is why technical associations publish journals. It also explains an important function of library research in technical curricula: to provide one of the skills necessary for self-education.

This chapter discusses the basic techniques for locating and using published information. In many ways, library research resembles laboratory research. A successful laboratory researcher knows his equipment, devises a plan for reaching his objective, and keeps careful records. This systematic, scientific approach, which is familiar to all technical students, works just as effectively in library research.

WHERE TO START

Upon entering the library, you confront thousands of books and periodicals, knowing that perhaps ten of them are potentially good sources.

To find these ten, immediately examine the card catalog and periodical indexes. The following paragraphs explain how to locate potentially useful publications in the catalog and indexes. Encyclopedias are also discussed because they refer you to pertinent books and periodicals.

ENCYCLOPEDIAS

Encyclopedias serve only as points of departure because they do not provide the type of detailed information needed in reports. However, they do give background information, and many of their articles contain bibliographies naming sources of more specific and useful data. The four most popular encyclopedias are described below, followed by a list of smaller, specialized encyclopedias.

Encyclopedia Americana
   Considered a good source for general technical information; supplemented yearly by the *American Annual.*
Encyclopaedia Britannica
   Considered the best general encyclopedia; supplemented annually by the *Britannica Book of the Year.*
Encyclopedia of Chemical Technology
   Covers all fields of chemical technology; fifteen volumes; now includes two supplemental volumes.
Encyclopedia of Science and Technology
   Covers major technological applications of all the natural sciences; fifteen volumes; supplemented yearly.
The Encyclopedia of Chemistry
The Encyclopedia of Chemical Process Equipment
The Encyclopedia of Electronics
The Encyclopedia of Management
The Encyclopedia of Physics

THE CARD CATALOG

The card catalog lists every book in the library. Many libraries list each book in three alphabetical files containing (1) author cards, (2) title cards, and (3) subject cards, as shown in Figure 1. Except for their main headings, the cards are exactly alike. Items on the subject card in Figure 1 are numbered for purposes of reference:

1. You begin by thinking of subjects under which pertinent information might be found. Then, examine the cards in the subject section of the card catalog. In Figure 1, the subject is "Population." Subject cards are heavily cross-referenced, so they often refer you to another subject. Also, some books are listed under several subjects, as will be seen in item 8 below.
2. The call number explains where the book can be found on the shelves. If the call number begins with letters, as in Figure 1, the library is organized according to the Library of Congress System. Where the Dewey system is used, this book's call number is 301.3, as specified at the bottom of this card.
3. The author's name appears here. To see if he has written other books which might be useful, check the author cards.
4. The name of the book appears below the author's name.

5. The place of publication, publisher, and year of publication are listed in that order. The year has importance if rapid advancements are being made in your subject area.

6. The collation gives the total pages of the book and, if applicable, indicates that it has illustrations or is part of a series. Other items, of little interest, might be the height of the book in centimeters, and the price.

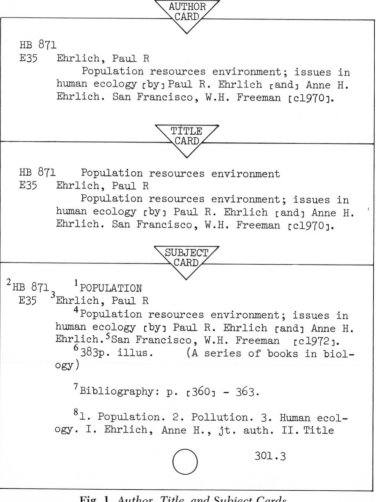

AUTHOR CARD

```
HB 871
  E35   Ehrlich, Paul R
          Population resources environment; issues in
        human ecology [by] Paul R. Ehrlich [and] Anne H.
        Ehrlich. San Francisco, W.H. Freeman [c1970].
```

TITLE CARD

```
HB 871    Population resources environment
  E35   Ehrlich, Paul R
          Population resources environment; issues in
        human ecology [by] Paul R. Ehrlich [and] Anne H.
        Ehrlich. San Francisco, W.H. Freeman [c1970].
```

SUBJECT CARD

```
²HB 871      ¹POPULATION
  E35   ³Ehrlich, Paul R
          ⁴Population resources environment; issues in
        human ecology [by] Paul R. Ehrlich [and] Anne H.
        Ehrlich.⁵San Francisco, W.H. Freeman  [c1972].
          ⁶383p. illus.    (A series of books in biol-
        ogy)

          ⁷Bibliography: p. [360] - 363.

          ⁸1. Population. 2. Pollution. 3. Human ecol-
        ogy. I. Ehrlich, Anne H., jt. auth. II. Title

                              301.3
```

Fig. 1 *Author, Title, and Subject Cards*

7. This section provides special information about the contents of a book, such as its inclusion of a bibliography. Bibliographies are extremely helpful because they lead you to additional sources of information.
8. Tracings name other subjects in the card catalog that list the book. Look under those subjects to insure that you cover all possible sources of information.

PERIODICAL INDEXES

Periodical indexes help you find magazine articles in the same way that card catalogs assist in locating books. Periodicals are published for all technical areas, and as mentioned earlier, they are especially good sources for keeping you informed of innovations in your field. Four major periodical indexes for technical subjects are listed below:

Applied Science and Technology Index
Subject index to articles in technology, engineering, science, and trade, including the areas of automation, construction, electronics, materials, telecommunication, and others.

Business Periodicals Index
Subject index to articles in all areas of business, including automation, labor, management, finance, marketing, public relations, communication, and others.

Engineering Index
Subject-author index to publications of industrial organizations, government, research institutes, and engineering organizations; includes annotations (descriptions) of the articles.

Reader's Guide to Periodical Literature
Subject-author index to articles on general subjects.

A list of articles appears beneath each subject heading in an index, as shown in the following excerpt from the *Applied Science and Technology Index*.

```
AIR pollution


Air pollution. D.A. O'Sullivan. il Chem & Eng N 48:38-41+
  Je 8 '70 (reprints 50¢)

Air pollution and human health. L.B. Lave and E.P. Seskin,
  Science 169:723-33 bibliog (p 731-3) Ag 21 '70

Air pollution by motor exhaust and other combustion pro-
  ducts. B. Szczeniowski. bibliog Eng J 53:4-10 Jl '70


Air pollution consultant guide/1970. map Air Pollution
  Control Assn J 20:495-9 Jl '70
```

Fig. 2 *Excerpt from a Periodical Index*

Subheadings appear beneath many subject headings to identify
more specific areas of information. The subheadings beneath "Air
Pollution," not shown in Figure 2, include "Air Pollution Laws and
Regulations," "Research," and "Testing."

Figure 3 deciphers the first entry in the excerpt above, an article
on pollution by Dermot O'Sullivan:

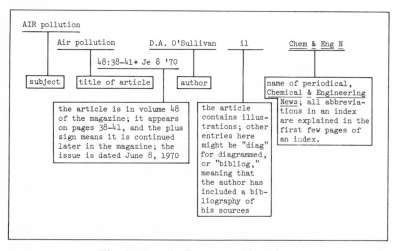

Fig. 3. *Periodical Entry Explained*

With this information, you need only the magazine's call number, found in the periodicals section of the card catalog, to complete a bibliography card for the source.

BIBLIOGRAPHY CARDS

As potential sources of information are found in the card catalog and periodical indexes, list them on separate three-by-five-inch cards. These bibliography cards should contain the name of the author, title of the article or book, and facts about its publication. You will use this information when documenting the sources of information used in the report. For your immediate use, write down the call number and any special information about the source. Below are sample bibliography cards:

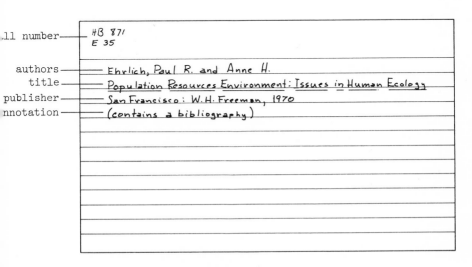

Fig. 4 *Bibliography Card for Book*

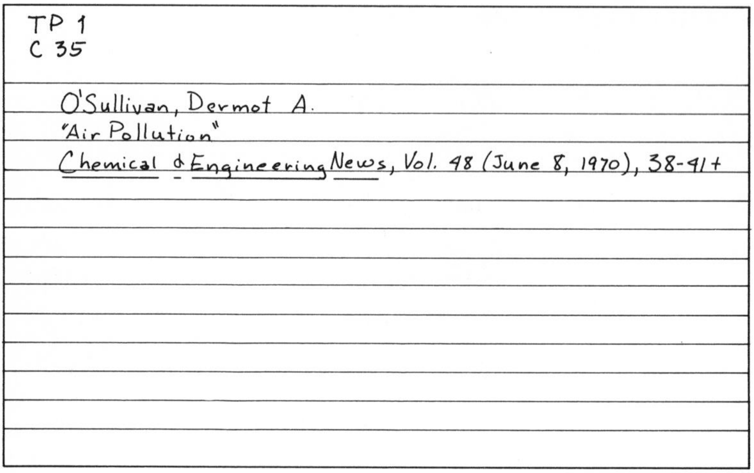

```
TP 1
C 35

O'Sullivan, Dermot A.
"Air Pollution"
Chemical & Engineering News, Vol. 48 (June 8, 1970), 38-41+
```

Fig. 5 *Bibliography Card for Periodical*

ABSTRACTS

After completing bibliography cards for all the sources, you can examine the appropriate abstracts to narrow your list of potentially good periodical articles. Abstracting journals, listed in the periodicals section of the card catalog, briefly describe articles in particular fields. The abstracts are generally descriptive rather than informative; that is, they state the scope of articles but do not summarize the articles. Therefore, they do not serve as substitutes for the article and cannot be quoted, but they do allow you to immediately reject inappropriate articles. The following is a brief list of technical abstracts:

Aeronautical Engineering Index
Chemical Abstracts
Engineering Abstracts
Geological Abstracts
Metallurgical Abstracts
Mineralogical Abstracts
Nuclear Science Abstracts
Science Abstracts (divided into sections on electrical engineering and physics)

TAKING NOTES

With your bibliography cards completed, go directly to the sources to begin reading and taking notes. Knowing the purpose and scope of your report, you already have a rough idea of its main headings. Place the appropriate heading, the author of the work, and the page number, on each note card. The headings may be altered somewhat during your research, but some uniformity will be maintained among the notes. Each card should contain notes on a single subject and from a single source. This seems like a waste of cards, but it greatly simplifies your card shuffling when you finally organize the report.

QUOTING AND PARAPHRASING

Quoting, or copying the words of the author verbatim, is done when a passage contains precisely the information needed for the report. Put quotation marks around each quote and footnote it, even if it contains only a couple of words. To shorten a quote, you

---

Cleanup

"Technology is responsible in large measure for pollution, so let's use technology to clean it up," is the stand that many enviromentalists take. They also frequently cite the fact that if we can land men on the moon and bring them safely back to earth again, we should be able to apply a like degree of American ingenuity and engineering skills to overcome the pollution problem.

There's no denying a certain degree of logic to these arguments. On the other hand, there is the very real question as to whether it will be possible, or indeed feasible, to mobilize the same degree of national effort to clean up the environment that it took to translate into reality President John F. Kennedy's vow to put man on the moon before the end of the '60's.

Nevertheless, efforts to reduce the emission of pollutants from various sources are rapidly gaining momentum. The results of such efforts may not be as dramatic as a space shot nor the results immediately evident, but at least a start has been made.

Fig. 6 *Original Passage*

can leave out some of the original words; mark such an omission, or ellipsis, with three spaced periods if the omitted words are in the middle of a sentence, and four spaced periods if they are at the end of the sentence. You can also insert short explanations into quoted material, but your own words must be enclosed by square brackets. These techniques are exemplified in Figure 7, a quotation from an article in *Chemical & Engineering News*.

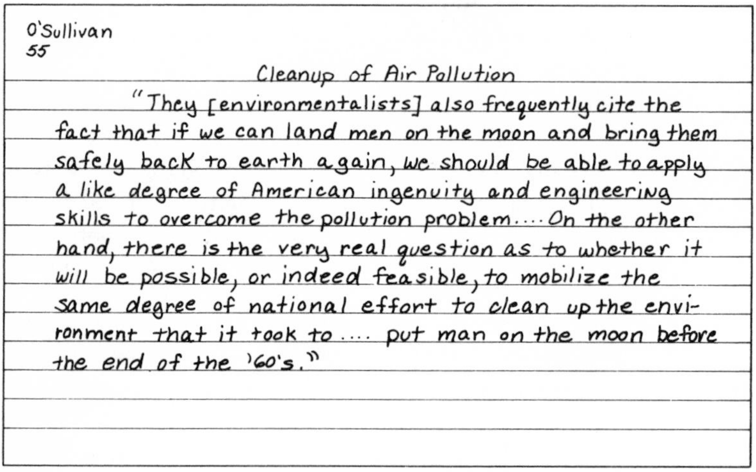

> O'Sullivan
> 55
>
> Cleanup of Air Pollution
> "They [environmentalists] also frequently cite the fact that if we can land men on the moon and bring them safely back to earth again, we should be able to apply a like degree of American ingenuity and engineering skills to overcome the pollution problem....On the other hand, there is the very real question as to whether it will be possible, or indeed feasible, to mobilize the same degree of national effort to clean up the environment that it took to .... put man on the moon before the end of the '60's."

Fig. 7 *Quotation from Passage*

When paraphrasing, you retain the meaning of a passage but put it in your own words. You cannot merely substitute synonyms and alter the author's sentence structure; if his statements are used that extensively, he must be quoted instead of paraphrased. Though not placed in quotation marks, paraphrases are footnoted. To avoid any doubt about where your ideas leave off and the paraphrased ideas begin, identify the author at the beginning of the paraphrase, as shown in Figure 8.

When quoting and paraphrasing, you have some obligations to both the original author and the report reader:

1. Whenever in doubt about whether an idea is yours or another author's, give credit to the other author.

2. Do not quote or paraphrase in a way that misrepresents the original author's meaning. The context of his words must not be distorted during the switch from a printed source to a report.
3. Never use extended paraphrases and rarely use extended quotes. If a report becomes nothing but other people's ideas thrown together, its objectives will not be met and the reader will have to conclude that you have done no independent thinking.

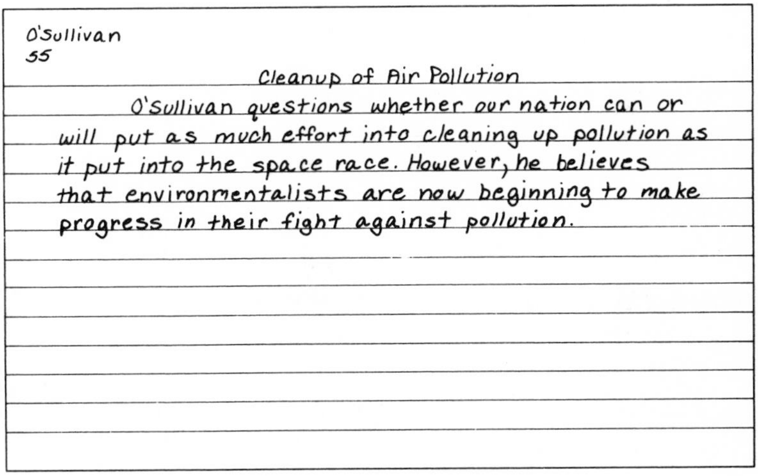

Fig. 8 *Paraphrase of Passage*

DOCUMENTING THE SOURCES

Within your report you must document, or give credit to, the sources of your information. Two common methods of documentation, (1) the footnote-bibliography format and (2) the author-year, references-cited format, are described below and exemplified in the two student models at the end of the chapter. Your instructor may elect to modify the formats.

FOOTNOTE-BIBLIOGRAPHY FORMAT

When using the footnote-bibliography format, place consecutive footnote numbers immediately after each quote or paraphrase. Then place a corresponding number and a statement of the source at the bottom of the page. Below are sample footnote forms for the most common sources of material; a complete list is available in the second edition of *The MLA Style Sheet* (May, 1970).

Footnotes can be used for things other than references to sources. If you place a definition or short explanation in a footnote, number it in the manner described above.

Reference to a book:

[1]Barry Commoner, Science and Survival (New York: Viking Press, 1966), p. 147.

[If the report contains a bibliography page, the name of the publisher can be omitted: (New York, 1966), p. 147.]

Reference to a book by two authors:

[2]Trevor Lewis and Lionel R. Taylor, Introduction to Experimental Ecology (New York: Academic Press, 1967), p. 230.

[The authors are named in the order they appear on the book's title page. If the book had more than two authors, the footnote would begin: Trevor Lewis and others, *Introduction*. . . .]

Reference to a second or subsequent edition of a book:

[3]Raymond F. Dasmann, Environmental Conservation, 2nd ed. (New York: Wiley, 1968), p. 112.

Reference to an essay in a volume of essays:

[4]Georg Borgstrom, "The Harvest of Seas: How Fruitful and for Whom?" in The Environmental Crisis: Man's Struggle to Live with Himself, ed. Harold W. Helfrich, Jr. (New Haven: Yale University Press, 1970), p. 82.

Reference to a work already cited:

[5]Commoner, p. 155.

[If two works by Commoner were being used, this footnote would also include the title or a shortened title of the book: Commoner, *Science and Survival*, p. 155.]

Reference to an article in a periodical:

[6]Dermot A. O'Sullivan, "Air Pollution," Chemical & Engineering News, 48 (June 8, 1970), 38.

[The "48" is the volume number. If no volume number was listed, the footnote would appear: . . . *News,* June 8, 1970, p. 38.]

Reference to an article in an encyclopedia:

[7]"Animal Ecology," Encyclopaedia Britannica, 1965.

[If the article were signed, the author's name would precede the article. Volume and page numbers are not needed for articles arranged alphabetically in reference books.]

Reference to a pamphlet or bulletin:

[8]L. N. Smith and W. L. Miller, Manpower Supply in Wastewater Treatment Plants, Technical Report No. 15, Purdue University Water Resources Research Center (Lafayette, Indiana, 1970), p. 3.

On your bibliography page at the end of the report, list the sources of material in alphabetical order according to the last name of the author. If the author is not named, list the article according to the first letter of the title, not counting "A," "An," or "The." A bibliography may contain all sources consulted or only those used, depending on the instructor. Each of the bibliographical entries below corresponds to a sample footnote above.

"Animal Ecology." Encyclopaedia Britannica, 1965.

Borgstrom, Georg. "The Harvest of Seas: How Fruitful and for Whom?" The Environmental Crisis: Man's Struggle to Live with Himself. Ed. Harold W. Helfrich, Jr. New Haven: Yale University Press, 1970, pp. 65-84.

Commoner, Barry. Science and Survival. New York: Viking Press, 1966.

Dasmann, Raymond F. Environmental Conservation. 2nd ed. New York: Wiley, 1968.

Lewis, Trevor, and Lionel R. Taylor. Introduction to Experimental Ecology. New York: Academic Press, 1967.

O'Sullivan, Dermot A.   "Air Pollution."   Chemical & Engineering
   News, 48 (June 8, 1970), 38-41.

Smith, L.N., and W. L. Miller.   Manpower Supply in Wastewater
   Treatment Plants.   Technical Report No. 15, Purdue University Water Resources Research Center.   Lafayette, Indiana,
   1970.

AUTHOR-YEAR, REFERENCES-CITED FORMAT

The author-year format, widely used in scientific writing, is shown
in the hypothetical examples below. When referring to an entire
work, place the name of the author and the year of publication in
parentheses:

An experiment (Markham, 1966) proved that the scanning beam

would work.

If the author's name is part of your text, place only the year in parentheses:

Markham's experiment (1966) proved that the scanning beam

would work.

When quoting, paraphrasing, or referring to a specific page or
pages of the work, include page numbers in the parentheses:

According to Markham (1966:34-35), "the scanning beam is

needed to control air traffic."

As the words suggest, a references-cited page lists only the
works cited within the report. The entries are listed alphabetically
according to the author's last name and contain essentially the
same information as bibliography entries. In references-cited entries, however, (1) the name of a book or periodical is generally not
underlined, (2) the name of a periodical is abbreviated, (3) the
title of a book or periodical article is usually not capitalized except
for the first word, (4) and the title of an article is not placed in
quotation marks. The following hypothetical entry for a book
and the references-cited page of the model entitled "Air Bags"

(p. 120) conform to the form used by the *American Journal of Science:*

```
Markham, James R., 1966, Control of air traffic:
   Chicago, Madison Press, 311 p.
```

These suggestions for documenting scientific papers will serve only as a point of departure. Before writing a term paper or thesis, find out the method of documentation preferred by your department. The department may provide you with its own style sheet, suggest that you follow the format in a particular periodical, or refer you to a book such as William G. Campbell's *Form and Style in Thesis Writing.*

The following student reports exemplify the two methods of documentation discussed in this chapter. The first model shows the footnote-bibliography format, and the second one uses the author-year, references-cited format.

THE ELECTRIC CAR AND ITS BATTERY

INTRODUCTION

The increased concern about air pollution in recent years
has emphasized the need for an alternative to the present internal-combustion engine. Nobody is certain that conventional
engines will be able to meet the requirements of the National
Air Quality Standards Act of 1970, but most consumers are
certain what will happen to automobile prices when Detroit
does meet the requirements.

This report will examine the electric car, one possibility
for economical, pollution-free travel, and will concentrate
on problems with the car's battery, which has always been its
greatest weakness.

BLASTING THE BATTERY

Problems with electric-car batteries were best exemplified in 1969 during a coast-to-coast race between cars
built by the Massachusetts Institute of Technology and the
California Institute of Technology. At the conclusion of
the race, E. H. Arctander, a journalist, wrote, "Don't hold
your breath waiting for an electric car you'd want to drive,
much less one you could afford."[1] His pessimism was based on
solid evidence: the cars, and especially their batteries, had
performed miserably.

[1]Erik H. Arctander, "What We're Learning About Electric
Cars," Popular Science, 194 (January, 1969), 106.

The battery problem is an old one with electric cars,
and many writers have explained how the battery overheats
during travel. In technical terms, the battery's power is an
inverse function of speed. More specifically, the faster the
car goes, the hotter the battery gets, and the sooner it becomes exhausted because heat causes resistance to current.
In addition, oxygen escapes from the heated electrolyte (an
electricity-conducting substance) as the battery discharges.
The oxygen forms an oxide layer on the battery's plates, not
only reducing power but making the battery difficult to recharge.

Arctander reported that the MIT team even risked blowing
up a battery to vaporize the oxide layer; the team charged
the battery beyond its full capacity (Arctander called it
"zapping"), and it worked, but it does not provide a practical
solution to the problem.[2]

Although not as a direct result of the car race, many
experts gave up on the electric car, and even its advocates
recognized that a breakthrough was necessary if the battery-
powered car was to become viable. One Cal Tech student summed
up his thoughts about the electric-car battery when he said,
"The trouble with electric cars is these crummy batteries. . . .
If I had the Baker motor and a good battery, I could beat anything."[3] The pessimism in this statement becomes obvious

[2]Arctander, p. 200.

[3]Arctander, p. 108.

when it is remembered that the Baker was an electric car that failed prior to World War I, just as the MIT and Cal Tech cars failed in 1969.

TRYING AGAIN

Good news for the electric car resulted from experiments not directly aimed at solving the car's battery problem. McCulloch Corporation, to improve its electric-powered chain saw, developed a method for recharging batteries that is applicable to electric cars. "[By applying] a series of strong charging pulses with a brief reversal of current, the McCulloch engineers found that they could dissipate the accumulated gases and successfully recharge the battery. 'The system,' says a McCulloch spokesman,'is comparable to the way a mother interrupts her feeding with burping to get rid of gas in the baby's stomach.' "[4]

An even more encouraging development is the Voltair, created by Robert R. Aronson's Electric Fuel Propulsion, Inc. In fact, the Voltair is viewed as a major breakthrough. The Voltair batteries "feature automatic recirculation of the electrolyte, allowing fast-charging in 20 minutes [without oxidation of the plates]. Used in combination with Aronson's patented fuel cell which trickle-charges the batteries

[4]"Technology: 'Burping' the Battery," *Time*, 94 (August 15, 1969), 73.

at a constant rate, the range of the Voltair will be at least 300-500 miles!"[5]

Voltairs have many other characteristics that make the car's acceleration and operating cost competitive with internal-combustion engines. When mass produced, the Voltair's initial price may also compete with Detroit's.

CONCLUSION

The electric car has failed so badly for so long that its recent breakthrough is viewed with cautious optimism. However, experts who were resigned to seeing only electric carts on golf courses instead of electric cars on expressways are greatly encouraged by the Voltair. It will take many years, but the electric car will eventually become an alternative to the polluting internal-combustion automobiles presently on the highways.

[5]Joseph P. Zmuda, "New Electrics Make Performance Breakthroughs," Popular Science, 198 (February, 1971), 56.

BIBLIOGRAPHY

Arctander, Erik H. "What We're Learning About Electric Cars."
  Popular Science, 194 (January, 1969), 106-109.

  Arctander emphasizes that internal-combustion engines not
  only pollute the air but also consume oxygen. He explains
  his discovery of a two-speed planetary transmission for
  the electric car and his early warning that the electric car
  needs a torque converter between its motor and its wheels.
  He concludes by saying that an electric car could be produced now if consumers would be satisfied with reduced
  performance.

"Technology: 'Burping' the Battery." Time, 94 (August 15,
  1969), 73.

  The article summarizes the problems with electric-car
  batteries and suggests that a battery-charging technique
  developed by McCulloch Corporation is applicable to electric cars.

Whiting, C. R. "Let's Produce Electric Cars Now." Science
  Digest, 70 (August, 1971), 68-71.

  Whiting describes the problems confronted by MIT and Cal
  Tech during a 1969 electric-car race. He says that electric-car experimenters need a real breakthrough if the
  car is ever to succeed.

Wouk, Victor. "Electric Cars: The Battery Problem." Science
  and Public Affairs: Bulletin of the Atomic Scientists, 27
  (April, 1971), 19-22.

  Wouk says there is little reason for believing that a
  battery-powered automobile can replace the conventional
  type within 30 years. He discusses the pollution problem
  and suggests that HEBAH (heat engine/battery hybrid) is a
  promising solution to the problem. The HEBAH is a combination engine-battery car which could be used until a
  battery-powered car is fully developed.

Zmuda, Joseph P. "New Electrics Make Performance Breakthroughs,"
  Popular Science, 198 (February, 1971), 55-56.

  Zmuda explains the dramatic advances made by Electric Fuel
  Propulsion, Inc., in developing the Voltair. He discusses
  the Voltair's battery system and describes how it operates.
  He concludes by relating the Voltair owner's plans for
  marketing the car.

AIR BAGS

INTRODUCTION

Despite the fact that safety belts are standard equipment on automobiles, over 50,000 Americans are killed each year in auto accidents and another 2 million are seriously injured. These numbers could be reduced by an estimated 40 percent if safety belts were worn, but very few people wear them. For the safety of occupants, it has become clear that autos need a passive protection system. "Passive" means that the system must require no effort by the passengers, and "protection" means that they must be safe during the equivalent of a frontal collision with a stationary barrier at 30 miles per hour. In 1968, Eaton, Yale & Towne Inc. announced an air-bag system that would meet these requirements.

OPERATION

Eaton's electro-mechanical system, which results in the rapid inflation of nylon air bags in front of an auto's occupants, is described below by Lund (1971: 64-65):

> Inflation is accomplished by a system of six components, the most essential of which is the sensor. Mounted on the car's firewall, the sensor must "decide" when the car has been in a collision equaling a barrier crash of eight miles an hour or more.
> /The/ sensor -- a little larger than a golf ball -- has a tiny spring that presses against a metal weight that must move forward about a half-inch, thereby completing an electrical circuit. . . .
> A couple of electric wires carry this "message" to a high-explosive cap on a bottle of nitrogen gas mounted behind the instrument panel. . . .
> When the cap explodes, the nitrogen rushes out into a "distributor," a sort of pipe with slits that quickly

distributes the nitrogen to a large coated-nylon bag
folded up in the instrument panel.

The Eaton system's two most impressive characteristics
are the sophistication of its sensor and the speed of the bag's
inflation. The sensor is able to distinguish between a collision and a noncollision. For example, the system would not be
activated if an auto knocked down a stop sign but <u>would</u> be
activated if, after hitting the stop sign, the auto rammed
into another auto. From the time of impact, activation of
the system and inflation of the air bag requires approximately
4 percent of 1 second, and deflation occurs almost as rapidly.
These features, however, are accompanied by several imperfec-
tions which have held up the bag's production.

PROBLEMS

Originally the federal government's Department of Transportation planned to require air bags as standard equipment
on all automobiles produced after January 1, 1973. The
deadline has now been set back at least 2 years. The main
problems, along with some possible solutions, are explained
below:

1. The system's sensor does not operate effectively
during angular collisions. According to Wargo (1970: 13),
Eaton will solve this problem by placing "two sensors. . .
at 30-deg. angles to the left and right of the car center-
line. The deceleration required to trigger the system in an
angular collision then would be only 87% of that required for
a frontal impact."

-2-

2. Air bags inflate with such tremendous force that a system completely safe for passengers has not yet been devised. Wargo (1970: 12) explains how this problem has led engineers in a vicious circle:

> It was discovered during tests with dummies that a child standing with his face close to the area where the bag emerged might be subjected to impacts in the "fatal" range. . . .
> The companies reduced the "standing child" problem by redesigning the diffuser system to deploy the air bag more vertically, but this change increased instrument-panel deformation. An excessively distorted panel could be a major hazard in multiple impact collisions, once the bag begins to deflate.

3. Coupled with the bag's potentially dangerous impact is the noise it makes during inflation. Lund (1971: 168) reports that the noise level has been successfully tested only on healthy adults: "Using a car with a small passenger compartment, Eaton's tests showed its system has about the same peak sound pressure as a shotgun. . . .But questions still remain about the ears of older people and motorists with ear trouble."

CONCLUSION

Auto manufacturers are now involved in perfecting air bags, and John Volpe, Secretary of the Department of Transportation, has postponed the deadline for installing them until at least 1975. However, Volpe seems to have become impatient (Williams, 1970: 23): "I know they work. I've seen them work. There is too much at stake. The problem now is for you (auto makers) to get the bugs out of the bags."

REFERENCES CITED

Lund, R., 1971, What ever happened to air bags? Pop. Mech.,
   v. 135, p. 63-65.

   Lund describes the operation of air bags. He also says
   that auto manufacturers are not dragging their feet in
   developing the bags. He states Detroit's main reasons
   for being skeptical of the bags, and lists Eaton's
   answers to these objections.

Wargo, J., 1970, Washington tells Detroit: cure auto accidents
   now: Prod. Eng., v. 41, p. 11-14.

   Wargo traces the development of air bags. He describes
   the bags being developed by Ford, General Motors, Chrysler, and Eaton, Yale & Towne, and explains the major
   problems with each of them.

Williams, D. N., 1970, "Air bag's got bugs," say automen:
   Iron Age, v. 205, p. 23.

   Williams explains the views of John Volpe, Secretary of
   the Department of Transportation, and provides General
   Motors' reasons for being unable to meet Volpe's deadline.

WRITING ASSIGNMENT

Write a 750-word research report explaining a recent innovation or development in your technical area to an uninformed reader. The introduction should provide background information about the development. In the conclusion, provide an estimate, preferably yours, of the significance of the development.

The body of the report can be organized in many ways. The following are some suggestions: 1) structure it according to problems and potential solutions regarding the development; (2) explain the development's applications in your technical area, in industry, or in society in general; (3) relate the development to general trends in business and industry. Your reader may have no rigid preconceptions about what should be included in the body of the report, but he will expect a logical, unified, and clear explanation of the development.

EXERCISE

Examine *Ulrich's International Periodicals Directory* or *The Standard Periodical Directory* in the library. They list the periodicals published for all fields. Select five of the most popular periodicals in your technical area and inspect several copies of each. Then write a one-paragraph evaluation of each periodical's potential usefulness to you following graduation.

7. ILLUSTRATING

The first four chapters on technical writing techniques concentrated on words but also mentioned illustrations. Illustrating is the technique of visually reinforcing report writing, particularly writing which involves descriptions and interpretations of statistics. Although they never act as substitutes for good writing, visuals should be included wherever they serve a function in a report.

Today's visual media have had a tremendous impact on everyone's communication skills. Mainly because of television, people grasp and interpret visual information more readily than printed information. Illustrations not only speed the communication process, which is important in our fast-paced industrial society, but serve as common denominators for report readers. Visual support accelerates an informed reader's rate of comprehension and helps insure the understanding of uninformed readers.

Another contribution of illustrations is the visual attractiveness they lend to reports. There is still a bit of the comic-book reader in most of us, and anything that contains illustrations simply looks understandable. For example, when we flip through a textbook for the first time, we usually make a judgment. If the book contains nothing but page after page of paragraphs, we suspect that it will be difficult to understand. If the book contains illustrations, we get the opposite impression, and that impression is accurate if the visuals actually clarify the text. Keep in mind that illustrations selected just for their attractiveness are like words chosen for their impressiveness. Words and visuals have only one purpose, to communicate information.

As a report writer, you select visual material by applying the same principles used for choosing words. Rather than insert the first visual that comes to mind, you consider the report's subject and reader. Then, taking advantage of the options discussed in this chapter, you decide which type of illustration will be successful. The ideas in this chapter are exemplified by visuals from the Heath Company's *Assembly Manual for the Citizen's Band Transceiver, Model GW-22.* Heathkit manuals are evidence that technical information, supplemented by effective illustrations, can be communicated successfully to readers who have varying levels of technical experience.

PHOTOGRAPHS

A photograph's advantage, the ability to show details clearly, can also be its disadvantage. At some points in a report too much detail can confuse the reader by detracting from what you want to emphasize. However, photography departments in industry are becoming very adept at arranging the subjects of photographs or airbrushing extraneous material out of them. Unless done skillfully, a photo is not inherently better than a drawing.

DRAWINGS

As indicated above, you have control over the content of your drawings and can therefore insure that they are integrated with the text of the report. Drawings are particularly helpful for readers who must become familiar with the physical characteristics of a mechanism before being able to understand its operation.

Having considered the reader in choosing an overall type of visual—drawings—you consider the object when deciding whether a cutaway, exploded, or cross-sectional drawing will work best.

CUTAWAY

A cutaway view removes a portion of the mechanism's casing. This shows the inside of the mechanism, revealing the inner parts' relationship to each other, and clarifying the position of the interior assembly in relation to its housing. Figure 1 exemplifies this, although a cutaway is more often used for smaller mechanisms.

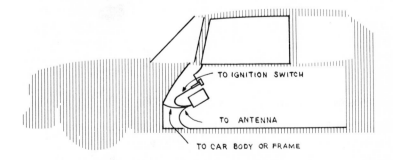

TO IGNITION SWITCH

TO ANTENNA

TO CAR BODY OR FRAME

Fig. 1 *Cutaway Drawing*

EXPLODED

As the word implies, an exploded view blows an object apart but maintains the arrangement of its parts, as in Figure 2. Exploded drawings are useful when you want to show the internal parts of a small and intricate object or explain how it is assembled.

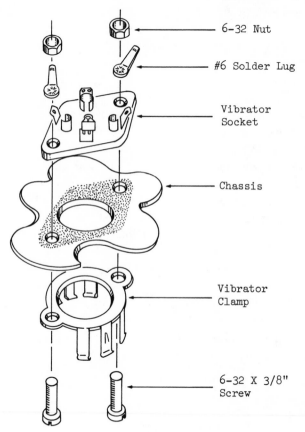

Fig. 2 *Exploded Drawing*

CROSS-SECTIONAL

Cross-sectional views are similar to cutaways except that they cut the entire assembled object, both exterior and interior, in half. In technical terms, the object is cut at right angles to its axis. A cross-sectional shows the size and relationship of all the parts. Two views of the same object, front and side views, for example, are often placed beside each other to give the reader additional perspective of the object.

PROCESS DRAWING

Another type of visual which emphasizes physical characteristics can arbitrarily be called a process drawing. It is generally reserved for manuals, where the directions sometimes need visual reinforcement. A step-by-step sequence of drawings assists the reader's performance of an operation such as soldering, shown in Figure 3. Operations involving only mechanisms, such as the strokes of a piston drawn on page 62 of this text, as well as processes including both man and machine, can be depicted either in simple drawings or photographs.

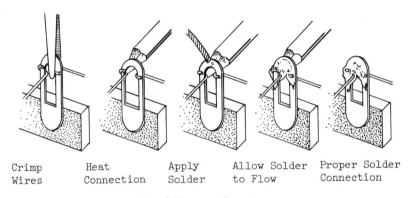

| Crimp Wires | Heat Connection | Apply Solder | Allow Solder to Flow | Proper Solder Connection |

Fig. 3 *Process Drawing*

DIAGRAMS

Although diagrams are discussed in a separate major section of this chapter, they are actually a type of drawing. The significant difference between the two is that diagrams communicate through symbols and do not try to show the physical characteristics of an object. This presents an immediate problem: is your reader well enough informed to understand the technical symbols? To achieve communication you must make the symbols conform to the reader's knowledge. This becomes a matter of selecting either a block diagram or a schematic diagram, because both show the operation of a mechanism.

BLOCK DIAGRAM

When a football coach draws a play on the blackboard, he uses symbols to indicate each player's assignment. In much the same way a block diagram represents several sequences or stages in a mechanism's operation. Each block has a label identifying the function of the sequence, and arrows indicate transition between the sequences. Make sure that the labels on a diagram are meaningful to your reader and that the diagram and paragraphs of description are well integrated. The circuit diagram in Figure 4, which represents the power supply sequences of the Heath Company's Citizen's Band Transceiver, helps clarify the manual's description of the operation.

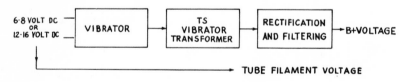

Fig. 4 *Block Diagram*

Block diagrams can also show the sequences of complex operations involving men and machines. Commonly called flow diagrams or flow charts, this type of diagram symbolically traces such processes as a factory's conversion of raw material into a finished product.

SCHEMATIC DIAGRAM

The power-supply process shown in the block diagram above is also represented in Figure 5 below, a schematic diagram. These two diagrams appear almost side by side in the Heathkit manual to serve all readers. The schematic symbols are aimed at an informed reader, but are clearly defined in the manual.

Schematics are used to represent systems in many fields, including electronics and hydraulics. Like blueprints, they are excellent time-saving devices if the reader understands them.

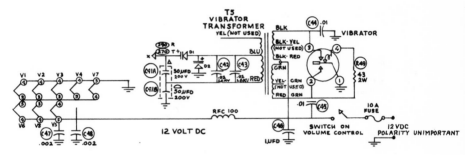

Fig. 5 *Schematic Diagram*

TABLES

Tables, which classify data in vertical and horizontal rows, are used extensively in reports to make statistical information readable. Pages 83–84 demonstrate their effectiveness by contrasting tabulated data with the same information presented in paragraph form. The student models in Chapters 5 and 11 provide additional examples.

Statistics are not the only type of data that can be tabulated. More and more, writers are categorizing and tabulating written information to speed the communication process. The troubleshooting table, or chart, from the Heath Company's Manual in Figure 6 exemplifies this. Technical writers, and manual writers in particular, should attempt to recognize types of information which can be simplified and emphasized by visual presentation.

DIFFICULTY	POSSIBLE CAUSE
Receiver section dead.	Check V1, V2, V3, V4, V7, and V8. Wiring error. Faulty speaker. Faulty receiver crystal. Crystal oscillator coil mistuned.
Receiver section weak.	Check V1, V2, and V3. Antenna, RF or IF coils mistuned. Faulty antenna or connecting cable.
Transmitter appears dead.	Check V5 and V6. Wiring error. Recheck oscillator, driver, and final tank coil tuning. Dummy load shorted or open.
Strong signal carrier from transmitter but no modulation	Faulty microphone connections, or microphone cable. Check V7 and V8. Wiring error.
Weak transmitted signal.	Faulty antenna. Shorted or open connecting cable. Poor antenna location.
Interference consistently encountered, either during transmission or reception.	Change operating frequency to more advantageous location in band. Check with others in your area for a frequency with minimum interference. After selection, order replacement crystals for your Transceiver from Heath Company. Give part number, frequency, channel number, and whether for transmit or receive.

Fig. 6 *Troubleshooting Table*

GRAPHS

Statistical information that lends itself to tables can often be given greater visual emphasis in graphs. Three common types are bar, pie, and line graphs. Many people call these charts, but either term is acceptable if it is used consistently in a report.

Bar and pie graphs are especially useful to the uninformed reader. In bar graphs, various bar lengths compare such things as production levels for each month of a quarter. Although a bar graph is highly visual, it is generally not as specific as a table, and should be used accordingly. Pie graphs present percentages visually, with the size of each slice representing, for example, the percentage of a project's total budget allotted to each project area.

A line graph, or curve, which allows statistical data to be plotted on vertical and horizontal axes, serves as another alternative to tables. Usually, independent variables such as years, stress, and frequency are plotted on the horizontal axis; the vertical axis indicates dependent variables such as profit, strain, and decibels. Depending on the information being presented, connection of the plotted points results in a single line showing a trend, or several lines emphasizing a comparison. Shading the areas between the lines results in a surface graph, which has additional visual emphasis. A line graph appears on page 197 of this text.

GUIDELINES FOR EFFECTIVE VISUALS

1. Integrate visuals into the report at logical and convenient places. A visual should not precede your discussion of it; if it does, the reader might spend fifteen minutes trying to understand the visual, only to discover that you have clarified it on the next page.
2. Refer to each visual in the text of the report, even if the visual is right beside or below your discussion of it.
3. Identify all visuals as either tables or figures; anything not a table is a figure, no matter what form it takes. Place numbers and titles directly above tables and directly below figures. Exception: If a report contains only one visual, do not number it.
4. Do not clutter a visual with too many words, causing it to lose

its visual quality and impact. When labels are needed for numerous parts, letter the parts, place the labels in a key next to the drawing, and use the letters for textual references to the parts, thereby saving space in both the visual and the text.

5. Do not assume that a visual appearing in a technical journal will automatically be effective in your report. As a report writer, your purpose, emphasis, and reader are often quite different from those of the technical article. If you do use a visual from a printed source, give credit for it in the lower right corner of the visual.

6. When your report contains an appendix, exercise your options by presenting highly technical visuals in the appendix and simplified forms of the visuals in the body of the report. Appendixes are discussed in Chapter 9.

7. Not only illustrations but anything different from the ordinary paragraph structure, such as a list or a heading, is visual because it calls attention to itself. Reserve emphasis for items which need it.

SECTION THREE

REPORT FORMATS

INFORMAL REPORTS AND MEMORANDUMS /
FORMAL REPORTS

SECTION THREE OVERVIEW

The techniques described in Section Two apply to both informal
and formal reports. However, the formats of these reports are
quite different. Section Three separates informal and formal
reports, explains why they must be approached differently, and
emphasizes methods of organizing them to communicate with
their respective readers.

8. INFORMAL REPORTS AND MEMORANDUMS

The day-to-day operation of a company depends on informal reports that circulate within and among its departments. These reports carry the results of investigations and convey information about products, methods and equipment. The most obvious difference between informal and formal reports is their format. Formal reports have title pages, tables of contents, and introductory summaries. Informal reports, whose formats will be emphasized in this chapter, generally have none of these; in fact, most informal reports are written on printed memorandum forms. Differences in the content of these two types of reports are outlined below:

Informal Reports		*Formal Reports*
attack either a problem of narrow scope or a limited number of aspects of a large problem	SCOPE	attack all aspects of a large problem
for departmental information or action, but occasionally to relay preliminary or partial results to management	PURPOSE	to play a role in a decision-making process at management level
are often aimed at only one person, who is usually technically informed	READERS	are read by several people, from as many management areas as are involved with the problem, some of whom fall into the technically uninformed category
emphasize the results of the investigation and the procedure by which data and results were obtained	EMPHASIS	emphasize the results of the investigation; the readers are interested in procedure only to the extent of knowing that the results are valid

The most significant of these differences is the readers' levels of technical sophistication. Invariably, some of a formal report's readers possess limited technical knowledge; therefore, the writer places highly technical information in the appendix of the report. On the other hand, an informal report's reader generally has a strong technical background. He is often an individual who has worked his way up to a supervisory position in a department. However, the informal-report writer cannot assume that his reader has total familiarity with the subject of the report. Every report involves investigation, thought, and writing, and the reader does not accompany the writer during all of these stages. If the writer has done his job well, he is inevitably more informed about the subject than his reader; keeping this in mind, he must present his information as clearly as possible.

TYPES OF INFORMAL REPORTS

All firms use printed memorandum forms for brief, handwritten notes whose verbal counterparts are messages conveyed by telephone. These notes, which conform to the strict definition of a memorandum as a written reminder, cannot properly be called reports. However, memorandum forms are so convenient that most companies use them to communicate the results of investigations. In other words, you will use memorandum forms to convey information which falls into the informal-reports category in the outline above.

The top of a printed memorandum form provides space for your name, the name of your reader, your subject, and the date. Depending on company practice, you may place the title and/or department of your reader as well as your own title and/or department in the space provided. The subject of your report, which is usually typed in capital letters, should be as specific as possible. Within the body of the report, use headings to emphasize its major sections, and reinforce your information with tables or figures. Place your signature or initials above your typed name at the end of the report. Formats and models of informal recommendation reports and laboratory reports, the two most important informal reports, are presented in the remainder of this chapter.

INFORMAL RECOMMENDATION REPORTS

The technique described in Chapter 5 for interpreting statistics applies to informal recommendation reports and formal feasibility reports (Chapter 11), which both require the evaluation of alternatives. As the writer of an informal recommendation report, your objective might be to select the better of two methods, or to determine which of three types of equipment your company should purchase. You begin by gathering information about the alternatives from printed material, observation, or interviews. Then you must select criteria by which to judge the alternatives. The most important criterion is generally "cost," but other factors such as "efficiency" or "capability" must sometimes be considered. An informal recommendation report's format appears below:

> Introduction
> > Purpose
> > Problem
> > Scope
>
> Body
> > Presentation and Interpretation
> > of Data
>
> Conclusions
>
> Recommendations

Throughout the report, you use an inductive approach. You take the reader through statements of the report's purpose, problem, and scope in the introduction. Then, after devoting the body of the report to presentation and discussion of data, you state your conclusions and recommendations.

This approach is exemplified in the following student report. The writer determines whether a proposed system would be more economical than the one presently used by the company. In the body of the report, he first demonstrates that the proposed system would reduce expenses. He then discusses the costs of installing the new system. Finally, he recommends purchase of the proposed system because it will quickly pay for itself.

MEMORANDUM

KAUZLARICH CONSTRUCTION COMPANY

TO: Mr. Dan Koeritz   DATE: September 19, 1972
Manager
FROM: Jim Biggs
Production Department

SUBJECT: RECOMMENDATION OF A METHOD FOR PURCHASING PAINT

INTRODUCTION

PURPOSE

The purpose of this report is to determine whether a
proposed method for purchasing paint would reduce costs.

PROBLEM

Because all steel strapping is painted to prevent corro-
sion, paint represents a major expense for the company. At
this time, the company purchases all its paint in 55-gallon
drums. It has been proposed that the paint be purchased by
tank truck at a saving of 10 cents per gallon.

The proposal would require installation of two permanent 650-gallon storage tanks at each of the company's two
high-volume machines. Also, twenty-eight 450-gallon portable
storage tanks would be required to maintain an adequate paint
inventory.

SCOPE

The only factor considered in this report is cost. The
report first compares annual paint costs under the present

and proposed systems. Then, expenses for purchasing and installing the proposed system are covered.

ANNUAL PAINT COSTS

The firm currently spends $2.18 per gallon of paint. Although the price of paint fluctuates, a 10-cent-per-gallon saving will always be given for tank-truck purchases. The two high-volume machines use 150,000 gallons annually. As Table 1 shows, this results in an annual cost of $327,000 under the present system as opposed to $312,000 using the proposed system. Thus, the proposed system would save $15,000 per year.

TABLE 1   ANNUAL PAINT COSTS

	Present	Proposed
Annual Usage	150,000	150,000
Cost Per Gallon	$2.18	$2.08
Annual Cost	$327,000	$312,000
Annual Saving		$15,000

COST OF EQUIPMENT AND INSTALLATION

Costs for equipment and installation will be considered only for the proposed system because these costs do not apply to the present system. The proposed system requires that a permanent tank be installed at each of the company's two high-volume machines. Also, 28 portable tanks are

-2-

needed to maintain a 1-month inventory. The prices for the tanks (Table 2) are $800 per permanent tank and $600 per portable tank, and each of the permanent tanks would cost $1,300 to install. The total cost for the proposed system would be $21,000.

TABLE 2   COST OF INSTALLATION AND EQUIPMENT

Permanent Tanks	Units	Total
Cost	2	$ 1,600
Installation	2	$ 2,600
Temporary Tanks	28	$16,800
Total Cost		$21,000

CONCLUSIONS

The proposed method of purchasing by tank truck would cut $15,000 from yearly paint costs. The cost of implementing this new system would be $21,000. Thus, the proposed system would operate at a $6,000 loss the first year, but would save $15,000 every year thereafter.

RECOMMENDATION

Because a $15,000 saving can be made annually after the first year, adoption of the proposed tank-truck system is recommended.

Jim Biggs

Jim Biggs

-3-

LABORATORY REPORTS

The function of a laboratory report is to communicate information gained through laboratory tests, which are the most rigid of all the methods of data-gathering. The accuracy of the test results depends on the procedure used during testing; therefore, the laboratory report places special emphasis on "apparatus" and "procedure," as the format below indicates:

Introduction
    Purpose
    Problem
    Scope
    Apparatus (or Equipment)
    Procedure
Body
    Presentation and Interpretation of Data (often called "Discussion")
Conclusions
Recommendations

A laboratory report's descriptions of apparatus and procedure affirm the accuracy of the data in the body of the report. Not all laboratory reports include recommendations, but the following student report attempts to determine whether a proposed concrete mix is strong enough to be used in a particular construction project. In technical terms, the specifications call for concrete that can withstand 5,000 pounds per square inch (psi); this specification, or criterion, receives a major heading in the body of the report. The writer does not include a scope statement because his statement of the problem clarifies the report's narrow scope.

MEMORANDUM

GARLITZ CEMENT COMPANY

TO: Mr. Jay Randall   DATE: August 5, 1972
     Manager, Testing Lab

FROM: Clinton Hare
       Senior Engineer

SUBJECT: TESTS ON CONCRETE FOR CANYON PROJECT

INTRODUCTION

PURPOSE

The purpose of this report is to present results of
tests on the concrete mix proposed for the Canyon Project
and to determine whether the mix is feasible.

PROBLEM

Specifications for the project state that the concrete,
which will be exposed to frequent freezing and thawing, must
be capable of withstanding 5,000 psi. The mix designed to meet
this requirement is shown in Table 1:

TABLE 1 DATA AND CALCULATIONS FOR MIX

Material	Weight (lb) Per Cubic Yard	Weight (lb) Per 28.7 Pounds
Cement	775	28.7
Water	340	12.6
Fine Aggregate	1,140	42.2
Coarse Aggregate	1,680	62.2
Air	--	--
Total	3,995	145.7

APPARATUS

1. Mixing tub
2. Five wax cylinders
3. Tinius Olsen testing maching (serial number 89377, capacity 400,000 lb)
4. Curing room

PROCEDURE

(1) One cubic foot of concrete was mixed according to the specifications above. (2) The concrete was poured into five wax cylinders and tamped three times during the pouring. (3) When the concrete had hardened, the forms were removed and the concrete cylinders were placed in the curing room to be moist-cured at 70 degrees Fahrenheit. (4) In the final stage of the procedure, a Tinius Olsen loading machine was used to test the cylinders under a compressive load. One cylinder was tested at 7 days, one at 14 days, one at 21 days, and the remaining two at 28 days.

COMPRESSIVE STRENGTH

As specified, the concrete used in the Canyon Project will be subjected to a compressive stress of 5,000 psi. During the test, however, the concrete ruptured at 3,650 psi. The designed and achieved stresses are shown in Table 2:

If the concrete was mixed and cured properly, it should have increased in strength throughout the 28-day period. The 7-day cylinder should have been near 70 percent of maximum strength, the 14- and 21-day cylinders should have shown gains, and the final two cylinders should have reached or surpassed 5,000 psi.

TABLE 2   RESULTS

DATE TESTED	DAYS CURED	MAX. LOAD (lb)	STRESS (psi)	DESIGNED STRESS
7/6/72	7	81,000	2,840	3,400+
7/13/72	14	89,000	3,130	3,750+
7/20/72	21	104,000	3,650	3,900+
7/27/72	28	97,000	3,400	5,000+
7/27/72	28	45,000	1,580	5,000+

Calculations for amounts of water, cement, and fine and coarse aggregates have been double-checked; no error was detected in the proposed design.

CONCLUSIONS

The most likely cause for the concrete's rupture at 3,650 psi is that during initial curing, prior to removal to the curing room, the cylinders were accidentally jarred, disrupting settlement of the concrete. Another possibility is that the cement was insufficiently tamped during the pouring process.

RECOMMENDATIONS

Because no problem can be found in the design, the tests should be repeated. The curing process should be closely watched, and great care should be taken during the tamping stages.

Clinton Hare

Clinton Hare

Like the informal recommendation report earlier in the chapter, this laboratory report has an inductive structure. It leads the reader step by step through the investigation, the reasoning, and, finally, the conclusions and recommendations.

In some companies, writers place the conclusions and recommendations at the beginning of the report immediately following a statement of the report's purpose. This deductive format, in which the supporting evidence follows the conclusions and recommendations, allows a busy reader to immediately see the results and decide for himself whether he wants to examine the data. A deductive format can be used for informal recommendation reports as well as laboratory reports.

An introductory summary may also be used in informal reports. The writer precedes the entire report with a concise summary of its sections. For example, the laboratory report could have been preceded by the following summary:

Tests were conducted to determine whether the concrete mix proposed for the Canyon Project meets the specified compressive strength of 5,000 psi.

The cement ruptured at 3,650 psi on the Tinius Olsen testing machine, but double-checking indicates that calculations for the design are accurate. Human error, either in insufficiently tamping the mix, or accidentally jarring the cylinders during initial curing, is probably responsible for the failure.

The mix should be retested, with particular care being taken during tamping and curing.

In about 85 words, the reader receives the gist of the entire report, including its conclusions and recommendations. He can then read the full report if he needs the details. Introductory summaries are emphasized in Chapter 9 because they are commonly used in formal reports.

Some companies use letters instead of memorandum forms to report important data to higher-echelon departments, so a letter report is exemplified at the end of Chapter 12. As the present chapter has shown, however, memorandum forms are by far the most versatile vehicles for communicating information within a company. They are used for reporting the results of investigations as well as for handwritten messages, and Chapter 9 shows how they also function as transmittal correspondence.

9. FORMAL REPORTS

As Chapter 8 suggests, formal reports have an important function in industry today. They provide the information management needs to make decisions affecting the future of departments or entire firms. To successfully perform its function in a company's decision-making process, a formal report must communicate with many people: executive and management personnel, senior engineers, perhaps legal and financial officers, and others whose areas will be affected by the decision. The technical knowledge of these people obviously varies tremendously, but the report must serve as the main source of information for all of them. The increasing need for quick and effective communication with a wide range of readers has caused changes in the format of formal reports:

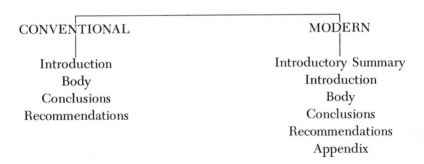

Transmittal Correspondence
Title Page
Table of Contents

CONVENTIONAL   MODERN

Introduction   Introductory Summary
Body   Introduction
Conclusions   Body
Recommendations   Conclusions
                                 Recommendations
                                    Appendix

The modern format includes an introductory summary and an appendix, two elements which greatly increase a report's flexibility. An introductory summary is the entire report in capsule form; it serves as a miniature report within the report. The appendix, which supplements information in the body of the report, contains highly technical information. With the addition of these two elements, a report adapts itself to all readers. By examining the introductory summary, busy readers quickly acquire the report's essential information; if they want more detailed information on a particular aspect, they inspect the appropriate section in the body of the report. Technically informed readers study the complex information in the appendix. The appendix allows the report writer to aim the body of the report at less informed readers.

In industry today, many firms have adopted the modern report structure, also called the administrative or double-report format. Figure 1 illustrates its effectiveness by indicating the number of readers who examine each of its parts:

```
SECTION
OF REPORT                FREQUENCY OF READING
```

Fig. 1 *How Managers Read Reports*

Richard W. Dodge, "What to Report," *Westinghouse Engineer*, 22 (July–Sept., 1962), p. 108.

The introductory summary receives the most attention because it summarizes the entire report. Its importance seems out of proportion with the time required to write it, but this is not really the case. In effect, a writer works on the summary while he prepares the body of the report; an introductory summary can hardly be better than the information it summarizes.

A complete formal report contains the following elements:

yes cover

yes Transmittal Correspondence — *single spaced*

yes Title Page

yes Table of Contents

no List of Illustrations

no Lists of Definitions and Symbols (if unavoidable)

yes Introductory Summary — *single spaced*

yes {
Introduction
Body
Conclusions
Recommendations
}

Bibliography (if applicable)

Appendix

The remainder of this chapter provides information about each of these elements and examples of most of them.

TRANSMITTAL CORRESPONDENCE

As the name implies, transmittal correspondence simply directs the report to someone. A memorandum (Fig. 2) conveys an intrafirm, or in-firm, report. An interfirm, or firm-to-firm, report requires a letter of transmittal (see Chapter 15, page 264). In either form, the information remains the same. The correspondence contains (1) the title of the report, (2) a statement of when it was requested, (3) a very general statement of the report's purpose and scope, (4) an explanation of problems encountered if, for example, some data were unavailable, and (5) acknowledgment of those who were particularly helpful in assembling the report. Transmittal correspondence should be only three or four paragraphs long because formal reports include an introductory summary, and saying too much in the letter or memorandum leads to repetition. The sample transmittal memorandum in Figure 2 (p. 148) is from a student report which will be used throughout this chapter.

Traditionally, transmittal correspondence has included a final paragraph in which the writer expresses hope that the report will be satisfactory. There is nothing wrong with this, but nothing particularly good about it either; it wastes words because it tells the reader something he already knows, and it simultaneously suggests the possibility that the report will not be satisfactory.

MEMORANDUM

N. P. SCOTT MANUFACTURING COMPANY

TO: Mr. Charles Tinkham   DATE: December 5, 1971
       Manager

FROM: Thomas E. McKain
       Engineering Department

SUBJECT: FEASIBILITY REPORT ON ADJUSTABLE-SPEED DRIVES

The attached report, entitled "The Feasibility of Adjustable-Speed Drives," is submitted in accordance with your request of October 15, 1971.

The report examines possible adjustable-speed drives to incorporate on the 3-1505 lathe. The cost and capability of a direct-current drive are compared to those for an alternating current drive.

The more feasible type of drive is recommended. Additional recommendations are made for purchasing the drive and establishing a maintenance program after its installation.

Thomas E. McKain

Thomas E. McKain

Fig. 2 *Transmittal Memorandum*

TITLE PAGE

Because the title page is among the reader's first impressions of the report, it should be well balanced and attractive. To achieve this, some firms have standard title pages just as they have letterhead stationery for business letters. As Figure 3 shows, title pages contain: (1) the report's title, (2) the name and title of the person to whom it is addressed and the name of his firm, (3) the name and title of the writer, and his firm's name if the report is interfirm, (4) and the date.

FEASIBILITY REPORT ON ADJUSTABLE-SPEED DRIVE

Prepared for
Mr. Charles Tinkham
Manager, N. P. Scott Company

By
Thomas E. McKain
Engineering Department

December 5, 1971

Fig. 3 *Title Page*

TABLE OF CONTENTS

A table of contents (Fig. 4) tells the reader what the report contains and where to find it, and also gives him an indication of the organization, depth, and emphasis of the report. Special-interest readers often glance at the table of contents, read the introductory summary, and turn to a particular section of the report. Others, after examining the conclusions and recommendations, locate certain data or calculations in the appendix.

To assist your readers, you must maintain uniformity between items in the table of contents and headings in the report. If, in the table of contents, a reader sees first- and second-level headings in capital letters, and third-level headings underlined, he should find precisely the same words and typographical devices in the headings of the report. Also, all items in the table of contents should be headings within the report, but not all headings in the report need to be in the table of contents. Two or three levels in the table of contents are sufficient to direct the reader, but lower-level headings may be needed within the report to achieve clarity.

Firms have various methods for numbering the pages of reports and placing items in the table of contents, but the following seems the most logical: Except for the title page, center lower-case Roman numerals (ii, iii) at the bottom of each page preceding the introduction. Starting with the first page of the introduction, center Arabic numerals at the bottom of each page. Include in the table of contents only those items following the table of contents. As Figure 4 shows, indentation should be used to help distinguish various levels of headings in the table. Roman numerals often precede first-level headings, but lower-level headings are not numbered.

LIST OF ILLUSTRATIONS

Illustrations refers to both tables and figures. Any illustrative material that is not a table is a figure, regardless of the way it is presented (drawing, graph, schematic, etc.). If illustrations appear only in the appendix, they can be listed in that section of the table of contents. However, illustrations usually appear throughout the

TABLE OF CONTENTS

Fig. 4 *Table of Contents*

body of the report as well as in the appendix; thus, they require a separate "List of Illustrations." The two types of illustrations should be separately numbered and listed as shown in Figure 5. Provide the number, title, and page for each illustration. The list of illustrations may be placed on the same page as the table of contents unless doing so crowds the page.

LIST OF ILLUSTRATIONS

Figures

    1. INITIAL COST GRAPH. 7

    2. EFFICIENCY CHART. 12

    3. AC BREAK-EVEN CHART 19

    4. DC BREAK-EVEN CHART 19

Tables

    1. COST COMPARISON OF AC AND DC DRIVES . . .11

    2. PROFIT DETERMINATION. 20

Fig. 5 *List of Illustrations*

LISTS OF DEFINITIONS AND SYMBOLS

Traditionally, reports have included lists of definitions and symbols; however, such lists are both unrealistic and unnecessary. Highly technical terminology and symbols should not appear in the body of a report. They should be reserved for the appendix, whose readers do not need definitions. When technical terms must be used in the body of the report, why not define them immediately in parentheses? Informed readers can simply skip over the definition.

If for some reason separate lists cannot be avoided, they should be detachable so readers can place them beside the text; few readers are able to quickly memorize and retain definitions of eight or ten unfamiliar terms. Symbols and definitions should be placed under separate headings, and definitions should be uniformly structured, although not necessarily in complete-sentence form.

INTRODUCTORY SUMMARY

The introductory summary (Fig. 6) functions as an independent unit, a mini-report. It represents the entire report, allowing busy readers to grasp the report's significant information without going any further. Most summaries include statements of (1) the report's purpose and the problem, (2) the major facts on which the conclusions are based, (3) the conclusions, and (4) the recommendations. In Figure 6, the first two paragraphs summarize the introduction of the report, paragraphs three and four summarize the body, and the last two paragraphs summarize the conclusions and recommendations. Try to concentrate this information into approximately 10 percent of the original report's length. Saying too much defeats the summary's purpose.

INTRODUCTORY SUMMARY

Summary of Introduction

The purpose of this report is to determine which adjustable-speed drive, ac (alternating current) or dc (direct current) should be purchased for production of the 19,500 steel crankshafts ordered by General Motors.

The number 3-1505 lathe must be converted to an adjustable-speed drive because its constant speed of 1620 revolutions per minute is too fast for making steel crankshafts. The ac and dc drives are compared according to cost and capability.

Summary of Body

Initial cost and installation of the dc drive is $122,-000, versus $102,500 for the ac. However, the operation and maintenance of the dc is approximately $14,000 less per year. With all costs subtracted from the production of the 19,000 units, the dc drive would show a profit of $203,000 (12.8

percent increase) versus $197,000 (9.8 percent increase) for
the ac.

The only difference in the capability of the drives is
their braking systems. The dc drive's system reduces the
input power required by the lathe, but the ac's is slightly
less efficient and costs about $3,000 per year more to
operate.

Summary of Conclusions
CONCLUSIONS

The ac and dc drives are both capable of doing the job,
and either would increase profits. However, the dc has a
3 percent greater profit potential than the ac.

Summary of Recommendations
RECOMMENDATIONS

The dc adjustable drive should be installed on the
number 3-1505 lathe. It should be purchased from the General
Electric Company, which has given assurance that it will be
installed within 2 weeks after purchase. With its implementation, a preventive maintenance program should be set up
to continually check the drive's components.

Fig. 6 *Introductory Summary*

A summary must be written after the rest of the report has been
completed. This seems repetitious, but the summary is not repetitious for readers who have no intention of looking at the rest of the
report. Technical terminology should be avoided because most
readers who really depend on an introductory summary are technically uninformed. Also, references to information in the body of the

report are unnecessary; the readers have access to this information through the table of contents.

Introductory summaries are somewhat like informative abstracts, which summarize a report's conclusions and recommendations. They are different from descriptive abstracts, which describe the contents of a report but not the results of the investigation.

INTRODUCTION

Flip Wilson's Geraldine says, "What you see is what you get," and that statement applies to report introductions as well as to Geraldine. An introduction prepares the reader to understand the body of the report and simultaneously keeps the reader from expecting something the report does not contain. Elements of an introduction are discussed below and exemplified in Figure 7.

PURPOSE STATEMENT

The first portion of an introduction, the purpose statement, can usually be expressed in one or two sentences. The best way to state the purpose is directly, by saying, "The purpose of this report is . . ." (to solve whatever problem necessitated the report).

```
                       INTRODUCTION


PURPOSE

     The purpose of this report is to recommend that either

an ac (alternating current) or dc (direct current) adjustable-

speed drive be purchased and installed on the number 3-1505

lathe for producing steel crankshafts.


PROBLEM

     The General Motors contract calls for 19,500 steel

crankshafts.  Number 3-1505 is the only lathe large enough
```

to produce crankshafts, but its constant speed of 1,620
revolutions per minute is good only for cutting softer
nodular-iron crankshafts.

SCOPE

    The ac and dc drives are compared according to their
cost (break-even point, first cost, rate of capital recovery), and capability (efficiency, overload capacity, braking, availability of parts). The sections on cost and
capability are preceded by background information on each
type of drive.

Fig. 7 *Introduction*

THE PROBLEM

Identification of the problem in the purpose statement leads you
into the introduction's second part, a definition of the problem.
Many people wonder why a definition is necessary, but it is important that the readers be completely informed about a problem before being told of its solution. During your investigation, you often
find that the problem is not what it first seemed to be; sometimes it
looks complex and turns out to be fairly simple, and often the reverse is true. Through your investigation of the problem, you become the expert. Your readers are comparatively uninformed, and
they will not understand your solution to the problem unless they
view the problem as you do. To supplement the definition of an extremely complex problem, you can devote the first section in the
body of the report to background information.

SCOPE

A scope statement reveals the emphasis, boundary, and organization of a report. In feasibility reports (see Chapter 12) the scope
section contains statements of the alternatives and criteria. In other

types of reports, you identify the main sections, or topics, of the report. By listing the sections in the order they appear in the body of the report, you also indicate the report's organization.

Given the type of introduction shown in Figure 7, the reader reaches the body of the report prepared to grasp its information.

BODY OF THE REPORT

The body of each type of report contains unique elements; each of these elements is discussed in subsequent chapters. However, all formal reports contain headings to clarify the relationship of the various elements and to serve as transitional devices. Headings also make reports more visually attractive and readable. Like other visual aids, they cannot be used for their own sake but should be included wherever logic permits.

First-level headings (INTRODUCTION, COST SECTION, RECOMMENDATIONS) are generally entirely capitalized and placed at the top of a new page (because of space limitations, these headings do not start a new page in the models in this book). Second-level headings are often entirely capitalized also, but as the examples below indicate, their placement varies. Third-level headings, with only the first letters of major words capitalized, are generally underlined and placed at the left margin to indicate their subordination. Fourth-level headings are capitalized and underlined like the third-level, but are indented five spaces on the same line as the first sentence of the new paragraph, and are followed by a period.

Two common methods for lettering and placing headings appear below. However, any logical system may be used provided the various headings appear uniformly throughout the body of the report and match their counterparts in the table of contents.

MAJOR HEADINGS	MAJOR HEADINGS
MAIN HEADINGS	MAIN HEADINGS
<u>Subheadings</u>	<u>Subheadings</u>
<u>Paragraph Headings.</u>	<u>Paragraph Headings.</u>

CONCLUSIONS

The Conclusions section emphasizes the report's most significant data and ideas. The readers must never be surprised by the conclusions, and they will not be if the report has been logically and clearly written. Although the information in the Conclusions section is extracted from the body of the report, it must be stated comprehensively, in terms of the overall problem. Concise, numbered conclusions, limited to information having the greatest impact on the recommendations, are the most effective (Fig. 8).

1. The initial and installation costs of the dc drive are greater than the ac's. However, the fixed and variable costs associated with the ac drive reduce the ac's total profit-producing capability. The dc drive would have a total profit of $203,000 for the 19,000 crankshafts, which is an increase of 12.8 percent over the present constant-speed drive. The ac drive would have an increase of only 9.8 percent.

2. The capability of the drives is very similar. The main difference between the two is in their braking systems. The dc drive has a regenerative brake which reduces the input power required by the lathe. The braking system of the ac drive is just as quick as the dc, but is not as efficient, and would cost as much as $3,000 per year more for input power.

Fig. 8 *Conclusions*

RECOMMENDATIONS

After viewing the conclusions, readers often know what the main recommendation will be. However, recommendations put you "on record," and should be as firm, clear, and concise as possible. The main recommendation usually fulfills the purpose of the report, but you should not hesitate to make further recommendations. In Figure 9, the writer suggests the initiation of a maintenance program to insure that the equipment he has recommended will perform well. If carried too far, this becomes "hedging," but done properly it serves both you and your firm.

1. Because of its profit-making potential, capability, and efficiency, the dc adjustable-speed drive is recommended for installation on the number 3-1505 lathe.

2. The drive should be purchased from the General Electric Company. They have given assurance that the drive can be installed in a maximum of 2 weeks.

3. A preventive maintenance program should be established by the machine shop. This maintenance program will force continual checking of the drive to see that its components are performing properly.

Fig. 9 *Recommendations*

BIBLIOGRAPHY

The Bibliography, which is included when the report contains information from other sources, is discussed along with footnotes in Chapter 6, "Researching Published Information." Many firms today have their own libraries which furnish useful information for formal reports.

APPENDIX

Writers frequently center the word *APPENDIX* on a fresh page to signal the start of this section. The appendix contains complex data, often in visual form, which supplement information presented earlier in the report. Today, the trend is toward greater use of appendixes to shorten the main report. However, you must not place so much in the appendix that you fail to present significant data in the main report; to avoid this, simplified versions of appendix items can often be prepared for the main report. In any case, refer to each appendix item at the appropriate place in the body of the report.

If the appendix has major sections, the headings should be lettered and placed in the table of contents (Fig. 4). Illustrations in the appendix must be given titles, numbered in the numbering sequence of the main report, and listed in the list of illustrations.

SECTION FOUR

TYPES OF FORMAL REPORTS

PROPOSALS / FEASIBILITY REPORTS /
PROGRESS REPORTS / MANUALS / ORAL REPORTS

Section Four is organized according to a sequence of report
writing that often occurs in industry. To oversimplify, an idea for a
change (new system, product, or service) is presented in a *proposal.* When several proposals are submitted, a *feasibility report*
may be written to determine which proposal should be implemented. As the change is being made, *progress reports* are written;
and after the change is completed, new *manuals* are often needed.
Oral reporting may reinforce or substitute for the proposal, feasibility, or progress reports in the sequence.

This overview serves as a point of departure for Section Four,
but it is not meant to suggest that each formal report you write
in industry will fit neatly into a sequence. The individual chapters
will explain the various situations that demand each type of
report.

10. PROPOSALS

A proposal presents a solution to a technical problem. There are two types of proposals, interfirm and intrafirm. The interfirm, or firm-to-firm, proposal responds to another firm's request for a solution to a problem. In an intrafirm, or in-firm, proposal, an employee or department conveys an idea for an improvement; often unsolicited, the report goes to someone in a higher position in the firm. These two types of proposals have similar content, but their functions are quite different, so they will be discussed separately in this chapter.

INTERFIRM PROPOSALS

In industry, the survival of a firm depends on its ability to write convincing interfirm proposals. These reports are the means by which a firm bids for and contracts. An abundant source for such contracts is the government, which requests proposals for innumerable items, the most familiar being national defense weapons. The following extract from *Parade* magazine's "Intelligence Report" (December 7, 1969, p. 12) describes the process by which proposals are solicited:

Last month the Pentagon asked five companies to submit proposals on major components of the B-1, a manned supersonic bomber designed to deliver nuclear weapons to enemy soil. The Pentagon plans to order 200 of these at an approximate cost of $25 million each.

The aircraft companies asked for proposals are General Dynamics of Fort Worth, Boeing of Seattle, and North American Rockwell of Los Angeles.

Pratt & Whitney (United Aircraft) of Hartford, and General Electric of Evendale, Ohio, were asked for engine proposals.

. . . The Air Force would like the B-1 to have a speed capability of 2000 mph, the ability to carry a goodly number of air to ground missiles, plus a heavy payload of nuclear bombs.

Along with each invitation, the Pentagon sent specifications stating requirements for the finished product. Specifications are extremely detailed and comprehensive, stating standards for even the most minute technical items, and specifying the content, format, and deadline for the proposal. Overall, the B-1's specifications demanded that it be able to reach 2,000 miles per hour and carry a certain number (the exact information is classified) of missiles and bombs.

When the Pentagon or any other large organization solicits proposals, it starts a chain reaction of proposal writing. The potential prime contractors for the B-1 (General Dynamics, Boeing, North American Rockwell), for example, have the ability to design and manufacture the airframe, but that is about all. Working from the original set of specifications, each firm writes its own specifications and solicits proposals from material suppliers, equipment manufacturers, and other subcontractors who can provide such items as radar equipment and hydraulic systems. Thus, firms which specialize in particular types of equipment submit proposals and compete with each other for subcontracts.

In an intense process that often takes several months, all of the firms involved attempt to write convincing proposals. As will be seen later, cost is only one of the criteria which determine who wins the contracts. The solicitor considers the quality of the product, financial status of the firm, qualifications of its personnel, and its "track record" for meeting deadlines. When the prime contractor finally selects its subcontractors, it compiles all of the data, submits a proposal to the Pentagon or whoever initiated the whole project, and hopes for the best.

The B-1 contract, won by North American Rockwell and General Electric, is worth a minimum of 5 billion dollars; thousands of workers, from laborers to engineers and management personnel, at Rockwell, GE, and their subcontractors are assured of jobs because of their firms' successful proposals. So many smaller firms depend upon Rockwell and other giants like it that proposal writing eventually affects a huge number of jobs. Sooner or later, most engineers become involved in the technical end of a proposal, and often the technical writing end.

Not all proposals are written for national defense items. Proposals are also common in state and local governments, public agencies, education, and industry. Competitive bidding through proposal writing is a basic part of our free-enterprise system.

ELEMENTS OF AN INTERFIRM PROPOSAL

The three main parts of a proposal are its technical, managerial, and financial sections, each of which amounts to a proposal in itself. None of these parts can be considered more important than the others. Cost has great importance, but the fact that the price is right becomes meaningless if the product is not, and both the price and the product become insignificant if a firm misses a deadline because of faulty management.

The difference between a winning and losing proposal is often an intangible: confidence. The tone of a proposal must be positive to convince readers that the company can be trusted to "produce" for them. Injecting confidence into a proposal presents more problems than might be readily apparent. The firm writing the proposal, like the firm requesting it, confronts the unknown. When its specifications were sent to bidders, the B-1 was nothing but an idea, some words and drawings on pieces of paper. So were its hydraulic systems and other components. To project into the future and convey a positive attitude, a writer must have confidence in his own company and his ability as a writer. The major parts of the proposal are described below.

Technical. A proposal's technical section begins by stating the problem to be solved. This seems unnecessary, but the firm must clearly demonstrate that it understands what the solicitor expects. Then, the firm describes its approach to the problem and presents a design for the product if one is needed. Sometimes, the firm offers alternative methods for solving the problem and invites the solicitor to select one.

Management. The management proposal outlines the chain of command that will be followed during work on the project. The solicitor wants to know that the project will be given top priority, and the proposal writer indicates that by explaining what positions and levels of management will be responsible for the success of the project. For added assurance, this portion of the proposal thoroughly describes how quality, costs, and schedules will be controlled. Finally, the firm includes a statement concerning its facilities, finances, and previous contracts.

Costs. The cost section provides a breakdown of the costs for every item in the proposal. Cost also involves the unknown because the firm must look into the future and estimate changes that may occur in the cost of labor, parts, and material during its work on the project.

After all the proposals are sent to the firm which requested them, they are studied by a team of evaluators, some of whom helped write the original specifications. First, the evaluators rate the proposals' technical and management sections without consideration of the cost data. A firm which submits an inadequate design or fails to effectively communicate its design loses the contract no matter how low its price. Then, evaluators examine the cost figures and select the best overall proposal.

INTRAFIRM PROPOSALS

All projects or changes begin somewhere, and that "somewhere" is usually on paper. To receive attention, an employee's idea generally has to be put in writing. Once on paper, the idea becomes an intrafirm proposal which will be evaluated by those in supervisory and management positions. Needless to say, departments have grown and careers have been launched by proposals. The individual who comes up with an idea for a project, and effectively communicates that idea, gains recognition and becomes a logical choice for a role in the implementation of the project.

An outline for an intrafirm proposal appears below, with the sections in the body of the report linked to their counterparts in the interfirm proposal:

Title Page
Table of Contents
List of Illustrations
Introductory Summary
Introduction
    Purpose
    Definition of the Problem
    Scope

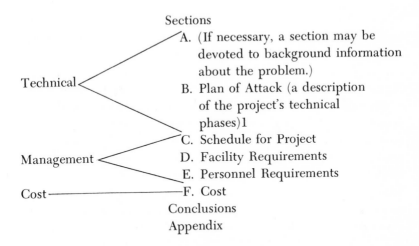

Sections
A. (If necessary, a section may be devoted to background information about the problem.)

Technical

B. Plan of Attack (a description of the project's technical phases)1

C. Schedule for Project

Management

D. Facility Requirements
E. Personnel Requirements

Cost ——————————F. Cost
Conclusions
Appendix

Chapter 9 describes a formal report's title page, table of contents, list of illustrations, introductory summary, and appendix. Therefore, the following discussion concentrates on the introduction and body of an intrafirm proposal.

INTRODUCTION

The introduction to an intrafirm proposal contains a statement of the report's purpose, a detailed explanation of the problem, and a concise statement of scope.

Purpose. A proposal's purpose is to present a solution to a particular problem. In stating the purpose, identify the problem but reserve amplification of it for the next section of the introduction.

Problem. The definition of the problem is the most important portion of a proposal's introduction. As the writer of an unsolicited proposal, you must demonstrate that a serious problem exists in order to justify the existence of your proposal. Even if the proposal has been solicited, your investigation should have given you greater insight into the problem than your readers possess. In either case, your definition must be clearly understood because the body of the report will attack the problem as you view it. As the outline above indicates, a section of background information may be placed in the body of the report. The section may include a detailed description of the existing situation (the problem) or a discussion of previous approaches to the problem.

Scope.   In the statement of scope, name the report's major sections in the order they will be presented. You thereby identify the boundary of the report and clarify its organization.

SECTIONS

The sections in the body of a proposal vary according to the type of problem being attacked; you must determine the content and structure that will best serve your proposal. However, the six sections explained below should at least be considered for inclusion in every intrafirm proposal.

Plan of Attack.   In the plan of attack, which is similar to the technical section of an interfirm proposal, analyze your proposed project, breaking it into phases involving equipment, or hardware; reserve the project's less technical aspects, such as the training of personnel, for the management section of the proposal. Implementation of any technical project, whether it will take eight hours or eight months to complete, can be logically divided into phases. For example, a writer proposing installation of a new conveyor system would describe each phase, from the dismantling of the old conveyor system to the installation of the new one. In developing this section, you must remember that unless your proposal has been solicited, your readers have no familiarity with it. The technical section requires clear writing if it is to be understood and accepted.

Schedule for the Project.   Always assume that your proposal will be accepted, anticipate difficulties that will arise during its implementation, and attack them. A work schedule is one of those difficulties. Eventually, someone must prepare a schedule for the project, and it might as well be you, particularly if you want to show how thoroughly you have planned the project. In the example above concerning a conveyor proposal, the writer should be capable of suggesting a reasonable date for the project to start, one which will allow its completion with minimum disruption of the firm's production schedule. He should also estimate the time required for each phase of the project. If the proposal includes a detailed schedule, several potential management headaches are avoided, and your proposal has a better chance for acceptance.

Facility Requirements. In intrafirm proposals, "facilities" generally refers to the physical capabilities of the firm. Availability of facilities plays an important part in proposals involving such things as the creation of a new product, enlargement of a department, or installation of a larger, more modern system. If the suggested change can be implemented without additional space, it is an obvious plus for the proposal. If, however, the proposal requires an expansion of present facilities, you must describe the physical alterations. Rather than state their cost in this section, wait until the cost section, where expenditures can be seen in their proper perspective. It is to be hoped that the detailed information which precedes the cost section will persuade the readers to accept the cost figures.

Personnel Requirements. Many proposals advocate the installation of new equipment. Such a report's personnel section can be divided into two sections, one concerning personnel necessary to complete the project, the other involving personnel needed to operate the equipment after installation. In the first portion, keep in mind that many firms today employ men with wide experience and diverse specialties; investigate your firm's potential and, if possible, suggest that the project be accomplished without costly outside help. The second portion deals with personnel required after completion of the project. If additional people must be hired to operate the equipment, this should be stated explicitly. If retraining the personnel is necessary, the length and type of training should be explained. Again, the costs for personnel are withheld until the cost section.

Cost. The cost section is the most important part of an intrafirm proposal. It contains an itemized list of prices for everything involved in the project, but, even more important, it explains how the money will be regained. In industry, an idea is as good as it is profitable; firms are usually able to find the money if they are assured that they can regain it and show a profit. Therefore, present clear and convincing evidence that your proposal has profit potential and should be acted upon. The "time-to-recover" concept discussed in Chapter 11 is one effective way to indicate eventual profit, but any logical presentation will be effective.

CONCLUSIONS

The conclusion of a proposal is essentially an inducement to action. There are three possible reasons that a problem has not been acted upon: people with the power to correct the problem (1) are not aware of the problem, (2) *are* aware of the problem but do not care, (3) are aware of the problem and *do* care but either do not know how to correct it or think that correcting it will cost too much. In the body of the report you have anticipated objections to your proposal and attacked them. In effect, you have attempted to convince the readers that solving the problem presents no problems. In the conclusion re-emphasize the advantages and strong points of your proposal.

The two student proposals which follow should give you some ideas for topics and for organizing your report. To save space in the bodies of the model reports, main headings do not each start a new page as they normally would.

PROPOSAL FOR
PROTECTION AGAINST SPREAD OF ALPHA RADIOACTIVITY

Prepared for
Mr. Donald Smith
Industrial Hygiene Department
Gawthrop Laboratories

By
Paul Markovich
Engineering Assistant, Plutonium Laboratory

October 29, 1970

TABLE OF CONTENTS

LIST OF ILLUSTRATIONS

INTRODUCTORY SUMMARY

Alpha radioactivity has often been spread over laboratory areas through the carelessness of people working in the area. Although detectors are available, they are not always used, causing the radioactivity on people to be inadvertently spread. By means of an electric-eye alarm spanning the laboratory's entrance, those who fail to use the radioactivity monitor will set off an alarm that can only be stopped when they use the monitor to check themselves. The alarm will warn personnel when they are entering as well as leaving the laboratory.

Our Electronics Division can adapt the present system to include an electric eye. Costs for this virtually "foolproof" system will total approximately $870, including a new monitoring unit and 1 year's maintenance on the entire system.

INTRODUCTION

PURPOSE

This report proposes a system for stopping the accidental spread of alpha radioactivity by people entering and leaving the plutonium laboratory.

PROBLEM

For years, radioactivity has occasionally been spread throughout laboratory areas which contain plutonium. There are various causes for these accidents, but the major problem seems to stem from individuals failing to monitor themselves when working in these areas. Presently, personnel are expected to always remember to check themselves on the alpha-radiation monitor, which is placed by the door of the laboratory. The inefficiency of this system has been proved on numerous occasions when alpha activity has been spread.

Alpha activity, which is the major hazard when working with plutonium, can go undetected and be spread quickly unless complete self-surveillance is undertaken. The chance of danger to the carrier, the shutdown of operations until the cleanup is complete, and the labor of reclamation crews make a more formidable monitoring system mandatory.

SCOPE

In this report, a description of the proposed system for monitoring alpha radiation is presented, followed by sections concerning equipment available, personnel available, costs, and finally, conclusions.

PROPOSED SYSTEM

This report proposes the installation of a "fool-proof" method of surveillance for all persons leaving or entering the laboratory. By means of an electric-eye alarm system spanning the entrance to the laboratory, all personnel will be reminded immediately that they are leaving or entering the area without checking themselves. If a person passes into or out of the laboratory without properly checking himself on the monitor, the electric eye will activate an alarm bell, reminding the person that he has forgotten to survey himself. He will have to use the monitor to deactivate the alarm system.

With the proposed system, a monitor will be placed immediately outside the laboratory door. As a worker depresses the probes on the monitor, a microswitch in the probe will de-energize the electric-eye system, which will be just beyond the monitor. The electric eye will remain de-energized for about 10 seconds to allow the person time to pass through its beam area. Once the 10 seconds are up, the electric-eye system will become activated again.

A person entering the laboratory will also be required to check himself. This precaution is taken so that radiation

from outside sources cannot be tracked into the laboratory.
A monitor will be set up on the outside of the electric-eye
beam, and anyone entering the laboratory without de-energiz-
ing the beam will also cause the alarm to sound. Figure 1
shows where the beam and monitors will be placed.

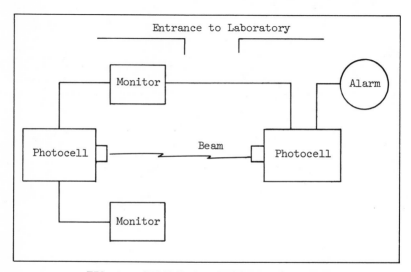

FIG. 1   DIAGRAM OF ELECTRIC-EYE SYSTEM

EQUIPMENT AVAILABLE

With the exception of one radiation monitor, which
will have to be purchased, the existing equipment for detection of activity can be quickly and cheaply modified to meet
the requirements of the proposed system. The major portion
of the electric-eye alarm system can be readily constructed
by our Electronics Division. The electronics staff can also

provide parts necessary for maintaining the monitors and electric-eye equipment. The new, transistorized units will consume less space than is required by the present system.

PERSONNEL AVAILABLE

Construction of the proposed monitoring and electric-eye alarm system can be undertaken by personnel in the Electronics Division, who are experienced in designing and miniaturizing this type of equipment. The mechanical and electrical equipment can also be maintained by electronics personnel. Periodic maintenance checks will be provided to insure optimum performance of the system at all times. If the system ever needs updating, the electronics staff will be able to make the necessary modifications.

COSTS

With the budgets in all laboratories somewhat limited, the costs for converting available equipment becomes important. However, the availability of most of the major components and the cooperation of the Electronics Division will hold the costs for this improvement to a minimum.

The major costs, at this time, appear to be for the new radiation monitor (approximately $350) and installation of the electric-eye equipment (approximately $240). Total

-4-

expenses for the system, including maintenance costs for 1 year, are estimated at $870. Table 1 contains a complete breakdown of estimated costs.

TABLE 1   COST ESTIMATES

Item	Cost
Equipment	
2 Photocells - - - - - - - - - - - - - - - - - - -	$100.00
1 Alpha Radiation Monitor- - - - - - - - - - - - -	350.00
Total- - - - - - - - - - - - - - - - - - -	$450.00
Installation Labor	
2 Men at $15.00/hour for 8 hours - - - - - - - - -	$240.00
Yearly Maintenance	
1 Man at $15.00/hour for 12 hours (1 hour/month)	$180.00
TOTAL ESTIMATED COSTS- - - - - - - - - - -	$870.00

CONCLUSIONS

1. The present method of checking alpha radiation on people entering and leaving the plutonium laboratory does not work satisfactorily. Protection of the individual has been left up to the individual, and by either forgetfulness or complacency, monitoring has sometimes been overlooked.

2. The proposed electric-eye alarm provides an almost "foolproof" system for monitoring personnel as they enter and

leave an "active" area. To turn off the alarm, personnel
must check themselves.

3. The safety feature of this alarm system would be a very
important asset to the plutonium facility. The safety of
individuals in the laboratories, and the elimination of
costly cleanup caused by the spread of radioactivity, make
this alarm system a necessity.

PROPOSAL FOR
INSTALLATION OF A POWER GENERATOR

Prepared for
David J. Boliker
Manager of Manufacturing
Lansing Steel Company

By
Robert J. Smith
Acting Project Engineer

January 15, 1971

TABLE OF CONTENTS

INTRODUCTORY SUMMARY

The purpose of this report is to propose a back-up generating plant at the new Lansing facility.

The problems which might result from an electrical power failure at the plant involve both personnel and facilities. Employee injuries could result from moving about with no lighting in the plant, being trapped inside closed tanks, and being struck by items falling from cranes which are operated by vacuum lifts or battery magnets. The plant itself would be in danger of damage from surface-drainage water, fire, and the freezing of pipes and heaters.

A surplus generator is available from our Field Force Division and could be installed for approximately $4,060, including labor and material. It can be installed in the powerhouse's air-compressor room during the month of June, and its operation will require minimum training of personnel. The generator can meet all the plant's emergency requirements, and do it very economically.

INTRODUCTION

PURPOSE

This report proposes the installation of an on-site electrical generating plant at the new manufacturing facility in Lansing.

PROBLEM

At present, a power failure would result in the following unsafe conditions for the employees: (1) plant lighting would be lost, and employees could not move safely within the plant; (2) employees working inside closed tanks could not get out until the tank could be rotated by an electric motor; (3) employees could be injured by items falling from the three overhead cranes which hold their loads by vacuum lifts or battery magnets and must set them down within 10 minutes after power failure.

A power failure would also cause damage to the plant: (1) the sump pump for surface-drainage water would not operate in the main office building and the basement of the gritblast area, causing extensive damage to equipment; (2) the fire alarm systems and the main fire pump would not function, and a fire could not be contained; (3) during winter, loss of the main heating boilers would cause water lines and fire-loop feedlines to freeze, break, flood areas of the plant, and ruin equipment.

SCOPE

In Section I, this report describes the installation of the generator. This is followed by sections on lighting and equipment changes, the schedule for the project, personnel training, and cost. At the end of the report, conclusions are presented.

SECTION I: INSTALLATION OF THE GENERATOR

A back-up generator is readily available for this project. One was purchased by the Field Force Division in 1963 for use as a standby unit at building sites. Because it was used only while other units were being repaired, the generator has operated for just 565 hours. Similar units have operated in excess of 15,000 hours, so the generator is almost new. It was built by Westinghouse and has a General Motors diesel engine.

The generator can be located in the air-compressor room of the powerhouse. This location will allow easy hookup to the main utility substation. The floor of the powerhouse is sufficient to hold the weight of the generator and to retain bolts needed to anchor the unit. The start-up fuel tank, located near the generator, will be used to operate the diesel engine.

Siding along the powerhouse's east wall must be removed to allow entrance of the generator; a louvered panel will

replace the siding and allow cooling air to circulate around
the generator. Installation of the generator will require
drilling holes in the floor to mount the anchor bolts. A
rack must be built to hold the batteries for starting the
diesel engine. An electric load-transfer panel will be
assembled and wired into the existing substation, and a fuel
line will be connected to the 2,000-gallon fuel tank. Hooking the generator to the substation requires installation of
three 460-volt cables between the powerhouse and the transformer. This can be accomplished by maintenance personnel.

SECTION II: LIGHTING AND EQUIPMENT CHANGES

In order to stay within the capacity of the generator
and still provide the power needed to maintain necessary
lighting and motor power, some electrical wiring changes must
be made in the plant. Lighting circuits now operated by
mechanical switches will have to be changed to electricity.
This change means that if the electric power fails, the
lighting will not turn on after auxiliary power is established,
thus removing electric loads during power failure. However,
the night lights, which will turn on automatically with the
start of auxiliary power, will allow employees to safely
leave the plant.

The rotoblast power supply will have to be changed to a
manual-start mode. This conversion requires the addition of

one relay to the panel board and allows personnel operating the cranes to lower suspended loads safely to the floor.

Sump pumps used for the removal of drainage water from the office building and basement of the gritblast will automatically return to operation when auxiliary power is supplied. No changes will be required in the existing circuits to these pumps.

The fire alarm will function as soon as power is transferred to the generator. If a fire should start, the alarm would be sounded immediately. At that time, all electrical power would be turned off, including the night-light circuitry. This is necessary because the fire-pump's motor taxes the generator so much that very little electricity is available.

SECTION III: SCHEDULE FOR THE PROJECT

PHASE ONE

Phase one will be completed during the normal work week, using plant personnel. This phase will include re-wiring the rotoblast, changing the lighting, installing anchor bolts, installing the fuel line, building and installing battery racks, and wiring the generator to the substation.

PHASE TWO

Phase two will involve Saturday work because two shifts of four men each will be required, and these men cannot be spared from duty during weekdays. Work will consist of removing the siding on the powerhouse wall, moving the generator into the powerhouse, installing it on its anchor bolts, and replacing the siding with a louvered panel.

PHASE THREE

Phase three will be accomplished on a Sunday when power can be turned off in the plant. Work will consist of hooking cables into the outside substation, hooking the fuel line to the engine, and connecting the power and battery cables to the generator. The power will be off for only one-half hour, and no special precautions need to be taken because no operating personnel will be in the plant.

Scheduling for this project can begin four weeks before the start of plant shutdown, which is scheduled to begin July 7. Phase one can start on June 12 and be completed by June 21. Phase two can be done on Saturday, June 22, and Phase three can be completed on Sunday, June 30.

SECTION IV: PERSONNEL TRAINING

No personnel training will be needed for starting the diesel generator. The control panel installed in the powerhouse will automatically start the engine in the event of a power failure and transfer the plant's reduced electrical load to the generator.

-5-

Operation of the fire pump from auxiliary power requires the manual operation of a start switch. All members of the plant fire team are familiar with this operation.

SUPERVISION

All shop supervisors will be instructed about the limitations of the auxiliary power. These foremen will tell their personnel which equipment should be operated to assure a safe shutdown. The foremen have enough experience in their areas to know what needs to be done. All instructions to supervisors will be covered by an interoffice memorandum when the auxiliary generator is operational.

MAINTENANCE

Routine maintenance on the engine and generator, using plant maintenance personnel who are familiar with this type of equipment, will be scheduled. The generator will be tested once a month to insure its availability for an emergency.

SECTION V: COST

The generator, which cost $14,657 in 1963, is now worth approximately $4,330, and is available at no cost.

Below are estimated costs for installing the generator, which include labor to build and install the necessary equipment, and costs for material.

Labor rates for plant personnel are based on the latest figures for this plant. Straight-time labor is $7.00 per hour, time and one-half is $9.25 per hour, and double-time is $11.50 per hour:

	Material	$2,450.00
	Labor	1,240.00
		$3,690.00
Contingency (10%)		370.00
TOTAL COST FOR PROJECT		$4,060.00

CONCLUSIONS

1. Our new manufacturing facility in Lansing represents an investment of $7 million and employs 300 people. Any interruption of electric power would create a total blackout and completely halt the equipment. This would result in very unsafe conditions for employees trying to move within the plant. Damage to equipment and spoilage of materials would also result if the power failure was prolonged.

2. Considering the safety of the employees, $4,060 for a back-up generator is an extremely reasonable investment. In addition, the facilities and equipment that might be damaged during a power failure would cost more than installation of the generator.

-7-

WRITING ASSIGNMENTS

1. Write a 1,000-word, problem-oriented formal proposal aimed at an uninformed reader. For your topic, either identify a real technical problem you are familiar with or create a hypothetical but realistic problem situation of your own. Define the problem clearly in your introduction. For the body of your report, consider each of the items in the outline early in this chapter. Base the proposal on factual data and make it convincing.

2. Write a 1,000-word formal proposal solving the problem described below:

The Midwest Oil Company is located near the shores of Lake Michigan. Three days ago, gale winds damaged five old storage tanks, and 35,000 gallons of crude oil were spilled within 24 hours. Repairmen were able to stop the flow from two tanks, but 7,000 gallons of oil per day continue to seep from the other three. Standard procedure would be to transfer the oil to other tanks but none are available; Midwest would never have used the old tanks if construction of new ones were on schedule.

The most pressing problem is what to do about the oil in Lake Michigan. The spilled oil has seeped into a canal next to the storage tanks, flowed into the lake, and spread over an area of 3 square miles. It now threatens a public beach adjacent to the oil refinery.

As environmental engineer at Midwest, you must propose a solution to this problem. After reading about the Oilevator in *Popular Science*, (as well as in *Time*, *Saturday Review*, and several other magazines), you decide that Midwest should purchase one to recover the oil:

Anti-Pollution Machine Laps Up Oil Slicks

Start its engine and the Oilevator's conveyor belt picks up oil floating on polluted water and lifts it to a wringer at the top of the machine. The wringer squeezes out the oil, which runs down into a recovery barrel.

Simple yet effective, the Oilevator can lap up 30 gallons of oil a minute, or 43,000 gallons a day. In trials, its harvest averaged 95 percent oil. Even in a swell, with the lower end a foot below the water's surface, pickup was 85 percent oil.

"One Oilevator, used with a containment boom, could have picked up the oil leaking off Santa Barbara (Calif.) relatively easily," Richard Sewell, the machine's inventor, says confidently. Sewell, who is scientific officer for the Canadian Defence Research Board in Esquimalt, B.C., began work on the machine five years ago. The Canadian Navy had asked for a device to clean up harbor oil spills, a serious fire hazard.

The Oilevator operates on the principle that oil and water don't mix and—an observation of Sewell—that nubby cotton holds a great deal of liquid for its weight. The conveyor belt—canvas faced with terry cloth—is first coated with the same kind of oil as it will pick up. When the belt starts moving, floating oil clings to the oil-wet terry cloth.

Oil companies and government agencies in the U.S. and Canada are interested in the machine, which a Canadian firm has begun making commercially. It sells for $7,500 through Bennett International Services Ltd. of Vancouver, B.C. Bennett offers a complete oil-pollution defense service: Oilevator, oil-containment boom, and trained crews. They're air-lifted to trouble spots.

Source: Clive Cocking, *Popular Science Monthly,* April 1970, p. 59. © 1970 Popular Science Publishing Co. Reprinted by permission.

In your proposal, stress the Oilevator's superiority over more conventional methods and explain how you intend to station it in the lake. Aim the report at an uninformed reader and address it to Mr. Samuel Tuckey, Vice President of Midwest Oil Company, Chadwick, Illinois.

11. FEASIBILITY REPORTS

In industry, change is a sign of stability; firms must change in order to remain competitive. Fierce competition with other firms demands that companies constantly create new products, modify old ones, and develop faster, more economical methods of producing their products. Because of the emphasis on innovation, change is not only a sign of stability, it is the most stable thing in industry; if firms can be certain of nothing else, they know that tomorrow will bring change.

Ideas for change are molded into intrafirm or interfirm proposals and submitted for evaluation. When one proposal is presented, the company must decide whether to act upon it; if several are presented, the best one must be selected. Separating the good ideas from the bad, and the practical ones from the spectacular, requires extremely thorough investigation. Few firms can afford to implement an idea that does not produce the expected results.

The function of judging proposed ideas and recommending whether they should be acted upon belongs to feasibility reports. In industry, "feasibility" implies more than "possibility"; a feasible idea must be supported by evidence that it will succeed. A department which wants to make expensive changes must often demonstrate the feasibility of its suggestions. For larger matters such as expansion, development of a new product, or purchase of a new system, a team investigates all aspects of the proposed change before making a recommendation.

SELECTING CRITERIA

Selection of criteria, the most important elements in a feasibility report, occurs during the gathering of data. Criteria are the standards by which the proposals are judged; therefore, they have tremendous impact on the final recommendation. Criteria vary according to the type of feasibility problem, as the following hypothetical situations indicate:

Problem: a city needs additional airport facilities
Alternatives: 1. expand the present airport
2. build a new airport at site A
3. build a new airport at site B

Criteria: 1. cost
        2. capability
           a. land and air space available
           b. access to highways and commercial centers
           c. ecological effects (air, water, noise
              pollution)

Problem: a firm needs a data-processing system
Alternatives: various-sized computer systems
Criteria: 1. cost
           a. initial
           b. operating
           c. maintenance
        2. capability
           a. to meet firm's present requirements
           b. to meet requirements if firm expands

Problem: the site for a branch plant must be selected
Alternatives: proposed sites in various regions of the country
Criteria: 1. cost
           a. construction
           b. transportation
           c. labor
           d. taxes
        2. capability of local utilities

Good feasibility reports include all pertinent criteria, but as the examples suggest, the criteria can often be grouped into the major categories of "cost" and "capability." These criteria do not apply to all feasibility problems, but an effort should always be made to limit the number of major criteria; doing so helps organize both the investigative and writing aspects of a feasibility study.

FEASIBILITY REPORT STRUCTURE

Because of the many aspects of feasibility problems and the large number of potential readers, feasibility reports invariably demand a formal report structure. The following format suffices for virtually any feasibility report:

Title Page
Table of Contents
Lists of Illustrations
Introductory Summary
Introduction
  Purpose
  Definition of the problem
  Scope
    Alternatives
    Criteria
Sections
  A. (If necessary, a section may be devoted to additional background and introductory information)
  B. Presentation and Interpretation of Data for the First Criterion
  C. Presentation and Interpretation of Data for the Second Criterion
  D, E, etc.—Same as above for other criteria
Conclusions
Recommendations
Appendix

Chapter 9 thoroughly describes the elements of a formal report. In fact, portions of a feasibility report appear throughout that chapter to exemplify the title page, table of contents, list of illustrations, introduction, conclusions and recommendations of a formal report. Therefore, this chapter concentrates on presentation and interpretation of data in the main body, or sections, of a feasibility report.

PRESENTING AND INTERPRETING DATA

The body of a feasibility report must be structured according to criteria, with each criterion receiving a major heading. Beneath each heading, first define the criterion, because it has merely been identified in the introduction to the report. Then, present and interpret data for all the alternatives, judging them according to the particular criterion. Repeat this process in each section of the report.

COST SECTION

In addition to initial costs, the cost section often includes estimated expenses for installation, operation, and maintenance. The report excerpted in Chapter 9 attempts to determine whether an ac (alternating-current) mechanism is more feasible than a dc (direct-current) mechanism for manufacturing crankshafts. The writer presents his cost data in the following table:

TABLE 1   COST DATA

ITEM	AC DRIVE	DC DRIVE
FIRST COSTS		
SPEED DRIVE	$76,000	$110,000
AIR DAMPER	1,500	----
TACHOMETER	2,000	----
INSTALLATION	23,000	12,000
TOTAL	$102,500	$122,000
OPERATIONAL COSTS		
MATERIAL AND LABOR	$522,175	$520,620
PRODUCTION LOSSES	325	130
TOTAL	$522,500	$520,750
MAINTENANCE COSTS		
PREVENTIVE	----	$1,950
REPAIR	$20,000	5,000
TOTAL	$20,000	$6,950

Interpreting the data in this table, the writer explains that the operational costs are based on the production of 6,500 crankshafts per year. He further explains that operational and maintenance costs, unlike first costs, are fixed; in other words, the firm would incur operational and maintenance costs yearly but would pay the first costs only once. He emphasizes that the ac drive's higher operational and maintenance costs would eventually exceed the dc drive's higher first costs. For this reason, he states that the dc unit has greater profit potential.

If your cost section or any other section in the body of your feasibility report is lengthy and complex, you can end it with a short summary. However, summaries can become repetitious and should be limited to sections whose complexity demands them.

CAPABILITY SECTION

Just as various cost criteria can be combined into one section of the report, the rest of the criteria often fall into the category of "capability." As the earlier hypothetical situations suggest, these criteria include items which are difficult to translate into accurate cost data, such as ecological effects or the ability of a system to meet requirements in the event of expansion. Noise pollution, which could result in bad public relations and a loss of business, must be taken into consideration during selection of an airport site despite the fact that no one knows how it might affect profits. Often, a system's ability to handle expansion must be examined even though no one can predict the extent of future expansion. Because such criteria lack accurate data, they must be placed in the capability section rather than the cost section of a report.

Many "capability" items do not lend themselves to presentation of data in tables or charts, usually because precise data are not available. In such situations, you must describe their potential effects as thoroughly as possible. When data are available, however, they can be presented in visual form. The chart on page 197 presents information about the efficiency of the two drive systems.

To interpret this data, the writer defines "efficiency" as the output of a system in comparison to the input power required by the system. Referring to the chart, he explains that efficiency is figured for 1,200 revolutions per minute, the speed required to produce crankshafts. He emphasizes that the dc's efficiency is 5 percent greater than the ac's at that speed. He concedes that the effect of efficiency upon cost is difficult to estimate but suggests that the less efficient ac system could cost as much as $3,000 more per year in wasted power.

COST AND CAPABILITY COMBINED

An alternative method for determining feasibility is to evaluate the alternatives according to "time to recover," a concept that is becoming increasingly common in industry. This method applies only when all the data, including capability data, are convertible into cost figures. It requires extremely careful projections into the fu-

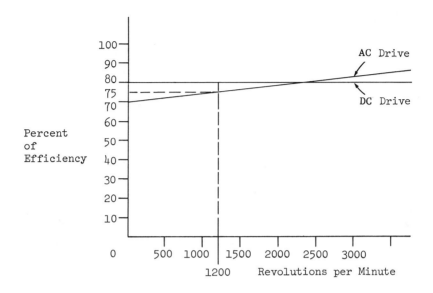

FIG. 1   EFFICIENCY OF AC AND DC DRIVE SYSTEMS

ture, and its result is a report with only one criterion, cost. The
process for figuring out time to recover is shown in the following
summary of a feasibility report.

The St. Louis warehouse of a steel company does not have coil-
processing equipment and can produce only flat sheet products. Its
annual sales are 2,260 tons, 1.3 percent of the area's total market
of 198,800 tons. The firm loses a large percentage of the market because approximately 75 percent of the sheet products sold in the
area are coiled rather than flat. The area's growing market is fore-
cast at 277,000 tons, but the warehouse will be able to compete for
only a small portion of it.

Faced with this situation, the steel company decides to study the
feasibility of installing coil-processing equipment. The final decision will be largely based on the time necessary for the firm to recover the money invested in the equipment, commonly called
"time to recover" or "years to recover." If the investment will pay
for itself within a reasonable time, it will be considered highly feasible, and the change will be made. Figures for the cost of the new
coil equipment are shown in the following table.

52-Inch Coil Slitting and Banding Line	$211,850
Installation	71,000
1 Jib Crane (6-ton)	8,500
Base	$291,350
Contingency (4.8%)	14,650
Total	$306,000

60-Inch Coil Cut-to-Length Line	$182,850
Installation	59,500
1 Jib Crane (6-ton)	8,500
Base	$250,850
Contingency (4.8%)	13,150
Total	$264,000
Grand Total	$570,000

The question now becomes "How soon can a $570,000 expenditure be recovered?" The answer necessitates thorough estimates of the potential market and projections of the firm's share of that market:

Added Sales - 15,600 Tons at $182/Ton		$2,839,200
Less Net Material	$2,189,900	
Selling Expense	348,800	
Property Tax	5,700	
Sales Tax	8,000	
Depreciation/Year	22,800	
Total	$2,575,200	
Net Profit before Federal Tax		$272,000
after Federal Tax		$124,000

Recovery of $570,000 capital at 3 year's penetration - 4.9 years

Although an entire "time-to-recover" evaluation includes additional items, the figures above show that the company expects, after three years of selling coil-processed steel ("penetration of the market"), to sell an additional 15,600 tons per year. Subtracting the expenses for material, marketing, taxes, and depreciation, the writer shows that the new equipment will earn $124,500 during its third year. The original $570,000 expenditure will be recovered in 4.9 years.

FEASIBILITY REPORT:

SITE FOR A CHICAGO SPORT COMPLEX

Prepared for
Mr. Cecil Roop, President
Roop Engineering Company

By
Robert W. Jonaitis
Chief Engineer

August 31, 1972

TABLE OF CONTENTS

LIST OF ILLUSTRATIONS

INTRODUCTORY SUMMARY

The purpose of this report is to examine the feasibility of three Chicago sites proposed for a new sport complex: Soldier Field, the Near West Side, and the South Loop.

City planners and engineers have determined that Chicago's present facilities, Soldier Field, Wrigley Field, and Comiskey Park, will soon become inadequate to serve the recreational needs of Chicago. The proposed sites are evaluated according to traffic accessibility and pre-construction costs.

Traffic accessibility is a problem involving the proximity of major expressways and their capabilities. The Soldier Field site has in the past caused major traffic problems, and the sport complex would multiply them. The Near West Side is adjacent to the Eisenhower and Dan Ryan Expressways, but this junction is already a major traffic problem during rush-hour traffic. The South Loop site is not as near to an expressway as the others, but with improvement of the roads between it and the Eisenhower Expressway, traffic could leave and enter the stadium area satisfactorily.

The total cost figures for each of the three sites, including purchase of land, razing the buildings, and improving traffic facilities, are as follows:

```
Soldier Field - - - - - - -$4,800,000
Near West Side- - - - - - -$3,500,000
South Loop- - - - - - - - -$5,800,000
```

The South Loop site is the most feasible for Chicago's new sport complex. Although its cost is highest, its central location will allow satisfactory movement of traffic. Modifications must be made in the exit from the Eisenhower Expressway at the north end of the proposed site.

INTRODUCTION

PURPOSE

The purpose of this report is to determine which of
three city locations would be the most practical and economical site for a sport complex. The alternatives are Soldier
Field, the Near West Side, and the South Loop.

PROBLEM

City planners and engineers have investigated and found
that the present facilities, Soldier Field, Wrigley Field,
and Comiskey Park, will be inadequate to handle the future
needs of Chicago. The cost of renovating the old stadiums
has been ruled out because of the high cost of remodeling
and the greater ability of a new site to handle the recreational needs of the entire city. Location of the complex
has been narrowed down to the three proposed sites.

SCOPE

Evaluation of the three sites is according to traffic
accessibility and pre-construction costs. Concerning traffic,
new cross-town expressways are presently being proposed, and
they could affect the choice of a stadium site. However, to
inject an uncertain variable into this study would only
reduce the reliability of the findings. Therefore, traffic
recommendations are based on the existing expressway system.

Section I examines traffic accessibility to the proposed
sites, and considers possible expressway modifications.

Section II presents costs for purchasing the sites, razing
the present buildings, and improving traffic accessibility.
Conclusions and recommendations follow.

SECTION I: TRAFFIC ACCESSIBILITY

An important factor to be considered for this sport
complex is how easy it will be for automobiles to enter and
leave the area. The seating capacity of the proposed stadium
is projected at approximately 60,000. Unless the surrounding
roads are adequate, an immense traffic jam will occur when
a capacity crowd enters and leaves the area.

SOLDIER FIELD

The Soldier Field site would create a traffic problem
because Lake Shore Drive is the only four-lane access road
in the area (Fig. 1). Although some traffic would enter from
directly west of the stadium, the bulk of traffic would use
the north-south route, Lake Shore Drive. The northbound
lanes would carry the heaviest load because many people would
be trying to enter the Eisenhower Expressway, which would
give them fast access to the far west suburbs of the city.
In the past, major traffic jams have resulted from cars
leaving the Soldier Field area after a sporting event, and
the new sport complex would result in even greater tie-ups.
Road improvements in the immediate area would only slightly
alleviate the back-up of traffic. Another factor to consider

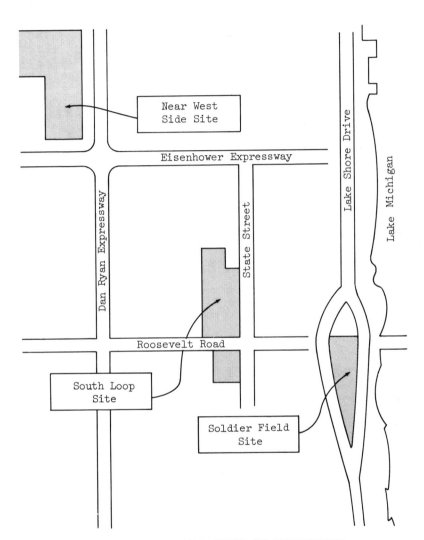

FIG. 1   PROPOSED SITES AND EXPRESSWAYS

is the possibility of additional traffic on Lake Shore Drive

resulting from events at McCormick Place.

For these reasons, the Soldier Field site appears to be

a poor choice for traffic accessibility. The flow of traffic

is limited to three directions because Lake Michigan is on
the east.

NEAR WEST SIDE

The second proposed site, on the Near West Side, has
the advantage, and disadvantage, of the Eisenhower and Dan
Ryan junction at its southeast corner. With the location of
the main parking lot at the junction of these two expressways,
access to and from the stadium, from all directions, would
appear to be very good. However, this intersection is
already very busy during the evening rush hours, and additional stadium traffic during this time would cause extensive
delays.

SOUTH LOOP

The South Loop site is near the center of the city, and
most people coming from the east and west parts of the city
would use the Eisenhower Expressway, which has an exit north
of the proposed stadium. People coming from the north and
south ends of the city would use either the Dan Ryan Expressway or Lake Shore Drive. From there they would get on the
Eisenhower and exit where it passes the stadium's north end.
State Street, which runs north and south, and Roosevelt Road,
which runs east and west, cross at the southeast edge of the
stadium site. These two streets would handle some of the
local traffic but most of it would bottleneck at the Eisenhower exit at the north end of the stadium. A modification

of this exit and convenient access to a north-stadium parking lot would ease the problem considerably. A south-stadium parking lot would be convenient for cars arriving via Roosevelt Road and State Street.

In summary, each of the three sites has a different type of traffic problem. Based on a careful projection of traffic patterns, the South Loop site has better potential than the other two. One of the most important concerns of the city is to avoid traffic tie-ups, and the South Loop, after modification of roads leading from the Eisenhower Expressway, would cause least congestion.

SECTION II: PRE-CONSTRUCTION COSTS

Before considering the cost of land at each site, it must be mentioned that the shape of each tract is different, a fact which will affect the design and cost of the complex. For example, the Soldier Field site is triangular in shape, and the design of a structure there could not be identical to designs for the other two sites. Although design and construction costs are beyond the scope of this report, the land tracts themselves will become a factor as construction of the complex is further studied.

SOLDIER FIELD

During this study, Soldier Field was examined by Roop Engineers, and the results of previous studies were confirmed:

the stadium is in extremely poor condition, and restoration would not only be almost as costly as a new complex but would cause serious limitations in the complex's design and potential.

The cost for razing Soldier Field would be $1,800,000, and the land is valued at $1 million.

An additional pre-construction consideration for Soldier Field is ecology. Park land would have to be used to provide additional parking space. A very strong campaign is now underway to preserve the lake front, and if this site is chosen there is no way to accurately predict what effect the campaign would have in terms of delays and ultimate costs.

NEAR WEST SIDE

The proposed site at the Near West Side is valued at $1 million and removal of old structures would cost approximately $1 million.

SOUTH LOOP

The South Loop site, because it is much nearer to the center of Chicago, is valued at $1,500,000. Razing of the many old buildings on this site would cost approximately $1,300,000.

Total costs for land, destruction of old structures, and improvement of traffic facilities at the respective sites are presented in Table 1 below. On a dollars-and-cents basis, the Near West Side site is $1,300,000 cheaper than Soldier

Field, and $2,300,000 cheaper than the South Loop.

TABLE 1   COST BREAKDOWN

	Near West Side	South Loop	Soldier Field
Removal of Old Structures	$1,000,000	$1,300,000	$1,800,000
Land	1,000,000	1,500,000	1,000,000
Improvement of Traffic Facilities	1,500,000	3,000,000	2,000,000
TOTAL COSTS	$3,500,000	$5,800,000	$4,800,000

CONCLUSIONS

The final decision on a Chicago sport complex site will

inevitably sacrifice either money or traffic convenience:

1.   Total expenses of the Soldier Field site are $4,800,000,

but its single large access road, Lake Shore Drive, severely

limits possibilities for satisfactory traffic flow.

2.   The Near West Side site has the lowest total price, $3,-

500,000, and is adjacent to the Eisenhower and Dan Ryan junc-

tion, but those expressways are already saturated with traffic,

particularly during evening rush hours.

3.   The South Loop site is centrally located and has the best potential for traffic accessibility. Because of its location, however, improvement of traffic facilities will cost $3 million, making its total price approximately $5,800,000.

RECOMMENDATIONS

1.   The South Loop site is the most feasible for Chicago's new sport complex.

2.   Its price tag exceeds those for Soldier Field and the Near West Side, but its central location allows comparatively easy access from four directions, making it worth the money to Chicagoans.

3.   The exit from the Eisenhower Expressway at the north end of the complex must be modified to avoid a bottleneck of traffic from State Street and Roosevelt Road.

FEASIBILITY REPORT:

LEASING OR PURCHASING A CRANE

Prepared for
Mr. George Strauch, President
Southern Construction Company

By
William S. Smolen
Assistant Manager

October 9, 1972

TABLE OF CONTENTS

LIST OF ILLUSTRATIONS

INTRODUCTORY SUMMARY

The purpose of this report is to determine whether it would be more feasible for Southern Construction Company to purchase an Anderson crane or to obtain a 10-year lease on one.

Southern will begin work on the steel superstructure of a 75-story office building in February of 1973. An Anderson crane is required to complete this project and fulfill contracts Southern hopes to win in the future. The feasibility of purchasing versus leasing the crane will be judged according to cost and requirements for supervision and maintenance.

The cost of leasing the crane would be $253,400. Borrowing money and purchasing the crane outright would cost $229,590, and an additional $2,000 would be required to train supervisory personnel. Thus, the total cost for purchasing would be $21,810 less than leasing. The schedule of payments also favors purchasing: purchase payments during the first 5 years would total approximately $106,000, versus approximately $230,000 during the same period if the crane were leased.

Because purchase of the Anderson crane would cost less and offer a better schedule of payments, it is more feasible than leasing the crane.

INTRODUCTION

PURPOSE

The purpose of this report is to determine whether Southern Construction Company should purchase or lease an Anderson crane.

PROBLEM

In September, Southern won a contract to construct the steel superstructure of a 75-story office building. The project, scheduled to start in February of 1973, is the largest the company has won. To complete this project and others Southern hopes to win contracts for, an Anderson crane must be acquired.

SCOPE

Southern has narrowed its alternatives to either purchasing the crane, which means borrowing the money from the First National Bank, or leasing the crane from Mitchell Leasing Company. The feasibility of these alternatives will be determined according to the criteria of (1) cost and (2) supervision and maintenance requirements.

COST

LEASING

Mitchell Leasing Company has proposed that Southern obtain a 10-year lease on an Anderson crane. For the first

-1-

5 years, Mitchell would charge $20 per month per $1,000 of

acquisition cost. The crane's acquisition cost is $400,000,

so rent would be $46,080 per year for the first 5 years.

Mitchell's fee is $2 per month per $1,000 of acquisition for

the second 5-year period, and payments would be $4,608 per

year. These figures take into account the 52 percent tax

shield covering the crane's rent, as shown in Table 1. The

total 10-year cost for renting the crane would be $253,440.

Ninety percent of the total cost would be paid during the

first 5 years ($46,080 per year) as shown in Table 4 in the

appendix of this report.

TABLE 1   LEASING COST

Years	Rent Per Year	Less Tax Shield (52%)	Annual After-Tax Cost
Fiscal 1974-78	$96,000	$49,920	$ 46,080
Fiscal 1979-83	9,600	4,992	4,608
			$ 50,688
			x 5 = $253,440

PURCHASING

 To purchase an Anderson crane outright, Southern would

have to borrow the $400,000 from the First National Bank.

The bank has proposed a 10-year loan bearing interest at 6

percent, payable monthly. The loan would be secured by a

mortgage on the heavy equipment Southern now owns. As shown

in Table 2, the net cost of purchasing the crane would be
$229,590. This cost takes into account the 52 percent tax
shield on interest, the 52 percent tax shield resulting from
the crane's depreciation, and the crane's book value after 10
years.

TABLE 2   PURCHASING COST

Price of Crane		$400,000
Total Interest	$132,000	
Less Tax Shield (52% of Interest)	68,640	
Net Interest Cost		63,360
Cost		$463,360
Less Depreciation Tax Shield (52% x $346,312 10-Year Depreciation)	180,082	
Less Book Value	53,688	
Net Cost		$229,590

Capital and interest payments on the bank loan are presented in Table 6 in the appendix. Table 5 shows the yearly
tax shield for depreciation and the yearly cost of purchasing
the crane. Greatest depreciation would occur during the
first few years of ownership; therefore, the crane's net
yearly cost would steadily rise from $9,920 in 1974 to $41,152
in 1983.

SUPERVISION AND MAINTENANCE

LEASING

Mitchell Leasing Company's fee covers the temporary services of a field engineer. These services are mandatory in
all of Mitchell's leasing agreements. The engineer supervises the crane's erection and the installation of additional
sections as construction proceeds. During the first year of
operation, he inspects periodically to see that the crane is
properly lubricated and that pulleys and cables are regularly
checked for wear. The engineer also supervises dismantling
when the first project is completed.

PURCHASING

Although several Southern employees are qualified to
operate the Anderson crane, none of them has experience in
erecting and dismantling it. However, two of Southern's
senior field engineers have directed these operations on
smaller cranes for more than 10 years and could quickly become familiar with the Anderson. Their training would take
approximately 1 month and cost a maximum of $2,000. Lubri-
cation and inspection of the crane is similar to that of
smaller cranes and could adequately be handled by Southern
personnel.

CONCLUSIONS

Table 3 shows that purchasing the crane would cost
$23,810 less than leasing it. When $2,000 for training supervisory personnel is subtracted from this figure, the difference
becomes $21,810. The schedule of payments also favors purchas-
ing the crane: leasing payments are $46,080 for each of the
first 5 years before dropping to $4,608 during the last 5;
purchasing payments rise gradually from $9,920 to $29,873
during the first 5 years and reach a maximum of $41,152 during
the tenth year. In all probability, Southern will be better
able to meet high payments during the second 5 years.

TABLE 3   SUMMARY OF COSTS

	Purchasing	Leasing
After-Tax Cost	$463,360	$253,400
Less Depreciation Tax Shield	180,082	----
	$283,278	$253,400
Book Value	53,688	----
Net Cost	$229,590	$253,400

RECOMMENDATIONS

Because of purchasing's lower total cost and favorable
schedule of payments, it is recommended that Southern borrow
$400,000 from the First National Bank to purchase the Ander-
son crane.

APPENDIX

TABLE 4 LEASING COST PER YEAR

	1974	1975	1976	1977	1978	1979	1980	1981	1982	1983
Net Rental Cost	46,080	46,080	46,080	46,080	46,080	4,608	4,608	4,608	4,608	4,608

TABLE 5 PURCHASING COST PER YEAR

	1974	1975	1976	1977	1978	1979	1980	1981	1982	1983
Payments and Net Interest	51,520	50,368	49,216	48,064	46,912	45,760	44,608	43,456	42,304	41,152
Less Depreciation Tax Shield (52% x Deprec.)	41,600	33,280	26,624	21,299	17,039	13,631	10,905	8,724	6,980	----
Yearly Cost	9,920	17,088	22,592	26,765	29,873	32,129	33,703	34,732	35,324	41,152

TABLE 6 BANK LOAN - CAPITAL AND INTEREST PAYMENTS

	1974	1975	1976	1977	1978	1979	1980	1981	1982	1983
Beginning Balance	400,000	360,000	320,000	280,000	240,000	200,000	160,000	120,000	80,000	40,000
Payment	40,000	40,000	40,000	40,000	40,000	40,000	40,000	40,000	40,000	40,000
Ending Balance	360,000	320,000	280,000	240,000	200,000	160,000	120,000	80,000	40,000	0
Interest										
6% on Ending Balance	21,600	19,200	16,800	14,400	12,000	9,600	7,200	4,800	2,400	0
6% on Payments	2,400	2,400	2,400	2,400	2,400	2,400	2,400	2,400	2,400	2,400
Total Interest	24,000	21,600	19,200	16,800	14,400	12,000	9,600	7,200	4,800	2,400
Less Tax Shield (52%)	12,480	11,232	9,984	8,736	7,488	6,240	4,992	3,744	2,496	1,248
Net Interest Cost	11,520	10,368	9,216	8,064	6,912	5,760	4,608	3,456	2,304	1,152
Payments and Net Interest	51,520	50,368	49,216	48,064	46,912	45,760	44,608	43,456	42,304	41,152

WRITING ASSIGNMENTS

1. Write a 1000-word problem-oriented formal feasibility report aimed at an uninformed reader. Your problem situation may be either real or hypothetical, but your data should be factual. Follow the outline presented in this chapter and evaluate two possible solutions to the problem. Clearly define the problem in your introduction and carefully select your criteria for evaluating the alternatives. Structure your report according to criteria rather than according to alternatives.

   The preceding student reports should help you organize your report. However, your problem situation should be more limited than the problem concerning the Chicago sport complex.

2. Write a 1,200-word formal feasibility report solving the problem described below:

   Auto Enterprises, a Los Angeles corporation, has been manufacturing automobile parts for 40 years. The corporation is now preparing to compete with GM, Chrysler, Ford, and American Motors by developing an economical, pollution-free compact car especially for metropolitan driving. Plans presently call for a prototype to be built by 1977 and for production of 30,000 compacts in 1982 at a retail cost of $4,100.

   Originally the firm intended to use a conventional internal combustion engine. However, the emergence of the Wankel engine has necessitated further study. As chief engineer for Auto Enterprises, you have been asked to determine whether the compact should be powered by a conventional engine or the Wankel.

   To write the report, you will want to gather information about the Wankel, examine recent developments in Detroit's present internal-combustion engine, and establish criteria by which to evaluate the two alternatives. You do not need to consider the cost of manufacturing rights to the engines. The following *Newsweek* article is one of many that have been written in recent years about the Wankel:

Autos: A Wankel for GM

When it first purred into the automotive picture sixteen years ago, the new engine was hailed as the first completely new development in its field in this century. By replacing the conventional piston with a triangular-shaped rotor within an elliptical combustion chamber, German scientist Felix Wankel had designed a power plant that was simple, compact and powerful—on paper, at least. But the leap from the drawing board to the production line turned out to be formidable, as Wankel's automotive partner, NSU, found out to its cost. In the event, ten years elapsed before the first commercial Wankel-powered car rolled out of NSU's factory. More than seventeen companies have purchased manufacturing rights, but only one of them—Japan's Toyo Kogyo—displays Wankels in its showroom, and sales have never been brisk.

For a time, it looked as if the Wankel would become just another auto-show curiosity—but no longer. The engine's future seemed secured last week when General Motors announced that it planned to purchase a production license from Audi NSU (now part of the Volkswagen combine), Wankel GmbH and Curtiss-Wright, which holds the North American rights.

Problems: If the deal is approved next month by the various corporate boards, GM will put up $5 million this year to iron out the engine's bugs. In all, the automaker will pay $50 million over the next five years to buy the right to produce and sell Wankel-powered cars anywhere in the world. Before then, however, several problems must still be solved. Partly because of poor lubrication, NSU's Ro80 has not earned high marks for reliability. One German auto magazine recently analyzed 191 of the cars driven an average of 20,000 miles and found that 65 per cent of them needed new engines.

But the most troublesome problem—preventing unburnt exhaust gases from escaping from the combustion chamber—appears to have been solved. Toyo Kogyo uses an aluminum and graphite compound to seal the tips of the rotors, while NSU favors a carbide coating. Either way, these innovations, coupled with conventional smog controls, enabled both Wankel cars to sail through a stringent Federal test for exhaust fumes. And the biggest attraction of all for GM is the prospect of substantial production savings. By switching to Wankel, which has relatively few moving parts, the automaker could eventually save as much as one-third of the cost of conventional power plants—an enormous dividend for a company making some 4.5 million auto engines each year.

Source: Newsweek, November 16, 1970, pp. 87–88. Copyright Newsweek, Inc. 1970, reprinted by permission.

Write your report for an uninformed reader and address it to the Board of Directors of Auto Enterprises Corporation in Los Angeles. If study in your field has given you a strong background on some other system for powering automobiles (electric, turbine, steam), your instructor may permit you to compare one of them to either the Wankel or the present Detroit engine.

12. PROGRESS REPORTS

Progress reports are written to inform management about the status of a project. Submitted regularly throughout the life of the project, they let the readers know whether work is progressing satisfactorily, which often means within the project's budget and time limitations. Contracts won through interfirm proposals invariably require that winning firms provide regular reports on progress to their clients. On the intrafirm level, departments furnish reports on a wide range of projects involving research, construction, and installation.

Progress reports are submitted at regular periods but differ from periodic reports such as weekly sales reports and monthly production reports. Unlike progress reports, periodic reports are continual, because the data they report never stops accumulating. Most firms provide standard, fill-in-the-blank forms to simplify the writing of periodic reports.

Also related to, but different from, progress reports are interim reports. The word *interim*, in this context, means "temporary." Workers near the completion of a project are often asked to report tentative results. They put together what amounts to a polished rough draft, presenting their data and stating tentative conclusions and recommendations. Management then uses the information until a final report is written. The main difference between progress and interim reports is their emphasis. When a progress-report writer reaches conclusions about some aspects of the project, he states them, but he generally does not include provisional results in each report.

INITIAL PROGRESS REPORTS

Preparation of a progress report forces you to view your project objectively and clarify your thoughts through writing. You must study the project to determine how things stand and state specifically what has been done and what needs to be done. The report speaks for the project, and if your writing lacks organization, the reader gets the impression that the project itself needs organization and direction.

The outline below applies to initial progress reports; subsequent reports are discussed later in this chapter. The outline can be adapted to virtually any type of project:

INTRODUCTION

    PURPOSE OF REPORT

    PURPOSE OF PROJECT

WORK COMPLETED (JULY 1 - JULY 31)

    INSTALLATION OF CUT-TO-LENGTH LINE

        <u>Conveyor System</u>

        <u>Crane</u>

    INSTALLATION OF BANDING LINE

        <u>Hydraulic System</u>

        <u>Expansion of Banding Area</u>

WORK SCHEDULED (AUGUST 1 - AUGUST 31)

    [same headings as for WORK COM-PLETED, although third-level headings (<u>Conveyor System</u>, etc.) change as tasks are completed and subsequent tasks are scheduled]

INTRODUCTION

In the introduction, remember that most readers are not informed about all aspects of the project. They might have walked through the laboratory or visited the construction site, but if they really understood everything they saw, a progress report would not be necessary. The introduction to an initial progress report requires particular care. To understand the progress being made and the problems that will be encountered, readers must fully grasp what the project involves.

Purpose of Report.   A single sentence usually identifies the type of report, names the project, and states the time period covered by the report. If subsequent progress reports will be written, state the number of the report, also.

Purpose of Project.   Having stated the report's purpose, you must spell out the entire project's objectives and scope; in interfirm project reports, you often quote directly from the contract that initiated the project. You then analyze the project, breaking it into major work areas. This analysis gives the reader a perspective of the project and prepares him for the body of the report.

WORK COMPLETED

The outline above, structured according to chronology and work areas, provides a logical organization for progress reports. Beneath the first main heading (WORK COMPLETED), second-level headings break the project into major tasks (INSTALLATION OF CUT-TO-LENGTH LINE, etc.). The scope of these major tasks should be comprehensive, allowing them to appear consistently in subsequent reports. On the other hand, third-level headings (Conveyor System, etc.) should be sequential, enabling you to replace them with other sub-tasks as the project moves forward.

WORK SCHEDULED

Under the heading for scheduled work, which also specifies the time period, second-level headings from the previous section appear again. They give the reader an opportunity to grasp the continuity of work in major project areas. Readers occasionally require a more detailed chronology of future work. To provide this, extend your format to include the following main headings:

WORK COMPLETED (DATES)

WORK SCHEDULED FOR NEXT PERIOD (DATES)

WORK PROPOSED FOR FUTURE (DATES)

  Lengthy progress reports often have a concluding section which summarizes the overall status of the project. However, instead of making a long report longer, consider preceding the entire report with an introductory summary.

SUBSEQUENT PROGRESS REPORTS

Second and succeeding progress reports maintain continuity and refresh the reader's memory by adding one section, a summary of work completed prior to the present reporting period:

INTRODUCTION

SUMMARY OF WORK PREVIOUSLY COMPLETED (DATES)

WORK COMPLETED (DATES)

WORK SCHEDULED (DATES)

The new section contains the same comprehensive headings for major tasks that have appeared in prior reports but condenses the information previously presented. To prepare this section, examine the "Work Completed" sections of your previous progress reports and write a capsule version of them.

You also shorten your introduction in subsequent progress reports. Change only the report's number in the "Purpose of Report" section but reduce the "Purpose of Project" statement to one or two sentences. This should be adequate for the readers, but they can always look in the files for the initial report's detailed description of the project.

SPECIAL PROBLEMS

Readers often indicate special interest in some aspect of the project that they particularly want to control. Budget, or any other item of special concern, can be given a main heading of its own and a thorough accounting.

A special section may also be necessary for requesting authority to change the project's scope; perhaps the investigation has produced an additional area that needs study. Before making a change that may affect the project's budget and deadline, put your recommendation in writing.

JONAITIS ENGINEERING COMPANY
1715 Mandel Road, Chicago, Illinois 60646

August 1, 1972

Mr. Cecil Roop, Chairman
Commission for Recreation
14 Randolph Street
Chicago, Illinois 60601

Dear Mr. Roop:

   Subject: Progress of Feasibility Study for
             Chicago Sport Complex (July 1 - July 31)

   This is a report of progress on the feasibility study
you requested. Soldier Field, the Near West Side, and the
South Loop are being studied to determine which would be
best for a sport complex. Criteria for the study are traffic
accessibility and pre-construction costs, which include land
value, cost of razing, and cost of road modifications.

WORK COMPLETED

TRAFFIC ACCESSIBILITY

   Research on potential traffic patterns has been completed for one of the three alternatives. The Soldier Field
site presents severe traffic problems. The only four-lane
access road in the area is Lake Shore Drive, and traffic
tie-ups for the large complex would be even worse than ones
which occur when Soldier Field is used. No road modifications would adequately solve the problem.
   Work on traffic patterns for the Near West and South
Loop sites began on September 15, and is still in preliminary stages.

PRE-CONSTRUCTION COSTS

   Except for the cost of traffic modifications at the Near
West and South Loop, all of the cost data have been gathered
and are presented in the following table.

Mr. Roop   -2-   August 1, 1972

COST-BREAKDOWN TABLE

	Near West Side	South Loop	Soldier Field
Land	$1,000,000	$1,500,000	$1,000,000
Removal of Old Structures	1,000,000	1,300,000	1,800,000
Improvement of Traffic Facilities	TO BE GATHERED	TO BE GATHERED	2,000,000

At this point, the Soldier Field site, which would cost a total of $4,800,000 and still not be adequate for reasons of traffic, has been tentatively determined unfeasible. Of the other two sites, the cost of land and razing at the Near West is $800,000 less than at the South Loop, but traffic data has not been gathered.

WORK SCHEDULED (AUGUST 1 - AUGUST 31)

TRAFFIC ACCESSIBILITY

Data for traffic patterns at the Near West and South Loop are being gathered and processing will begin on approximately October 15.

PRE-CONSTRUCTION COSTS

Figures for Soldier Field are complete, and cost data for the other sites are complete except for the cost of road modifications. Those data will be processed by November 1, when work on a draft of the final report will begin.

At present, this feasibility study is ahead of schedule, and a draft of the final report should be available prior to August 31.

Sincerely yours,

Robert W. Jonaitis

WRITING ASSIGNMENT

Your instructor may require a progress report approximately midway through your writing of a formal feasibility or proposal report. The progress report should be 350 words in length, aimed at an uninformed reader. The preceding interfirm progress report, written in letter form, describes progress on the feasibility study that appears at the end of Chapter 11. Notice the continuity between the "Work Completed" and "Work Scheduled" sections, and the writer's use of headings to clarify the report's organization.

13. MANUALS

Early in this text, technical writing was defined as writing that defies misunderstanding, and that statement is particularly applicable to the manual. The challenge to the manual writer increases as companies manufacture more complex equipment. Every time a firm submits a proposal and wins a contract to produce a new mechanism, a new manual is needed; in fact, the contract generally contains specifications covering the contents of the manual. Both the buyer and the manufacturer want the mechanism's success insured by precise directions for operation and maintenance.

As a manual writer, you obviously need to understand the mechanism, and you must also be aware of your readers' level of technical knowledge. You have no control over who will use the manual; unless informed otherwise, you must assume that the readers have limited technical backgrounds, especially the operators. Just as technical information in formal reports is aimed "down" to achieve communication, manuals should avoid highly technical terminology, symbols, abbreviations, and mathematics wherever possible. If some readers happen to be technically informed, and capable of understanding a more condensed and technical presentation, more power to them. In the manual writer's list of priorities, concern over insulting a reader's intelligence is very close to the bottom.

To assist the reader, drawings and diagrams appear throughout any good manual, providing heavy reinforcement for its descriptions and directions. Early in your planning of the manual, make tentative decisions about visual materials and work toward integrating them into the manual. Chapter 8, which includes examples from a Heathkit assembly manual, provides detailed information about the effective use of illustrations.

Manual writers also use headings and a rigid paragraph-numbering system to facilitate communication. After initially studying a manual, the reader uses it as a reference tool, and either of the following methods, combined with a table of contents, allows him rapid access to the information he needs:

1.0 SECTION	1–1 SECTION
1.1 COMPONENT	1–2 COMPONENT
1.1.1 Subpart	1–3 Subpart
1.1.2 Subpart	1–4 Subpart

1.2 COMPONENT   1–5 COMPONENT

   1.2.1 <u>Subpart</u>   1–6 <u>Subpart</u>

   1.2.2 <u>Subpart</u>   1–7 <u>Subpart</u>

The Dewey system on the left is extremely thorough; the digit-dash-digit system on the right depends on indentation and typographical devices to indicate subordination. Most manuals use these systems or some variation of them. Any system is fine as long as it provides uniformity and clear subordination.

TYPES OF MANUALS

In addition to training manuals, which are often indirectly related to mechanisms, the most common types of technical manuals, and their functions, are the following:

> Assembly: constructing, aligning, testing, and adjusting the mechanism
>
> Operation: operating it
>
> Service: keeping it operational through routine maintenance, such as lubrication
>
> Maintenance: locating malfunctions, testing parts, and repairing or replacing them

Extremely complex systems have separate manuals for these procedures, and sometimes several volumes for each of them. For smaller mechanisms, however, a single manual contains the information; this comprehensive, self-contained type of manual will be emphasized in this chapter.

SELF-CONTAINED MANUALS

In a self-contained manual, each major section builds upon the previous ones. First, you describe the mechanism's physical characteristics and explain its operating principles. Then, having fully informed the reader about the mechanism, you provide directions for operating it. The structure of a self-contained manual appears on the following page.

Title Page
Table of Contents
List of Illustrations
Introduction
General Description
Detailed Description
Theory of Operation
Operation
Maintenance
Parts List

The title page, table of contents, and list of illustrations are similar to those of a formal report. At the end of this chapter, a student manual on a surveying instrument exemplifies these and other elements of a manual.

INTRODUCTION

The introduction, which states the manual's purpose and scope, serves more as an introduction to the manual than to the mechanism. After stating the manufacturer's name and number for the piece of equipment, you explain the manual's functions, which are generally to supply operating and maintenance procedures. You then name the main sections of the manual in order, letting the reader know how it is organized. When applicable, call attention to other manuals or publications which provide additional information about the mechanism. The remainder of the introduction deals with any special background information necessary to familiarize the reader with the equipment.

GENERAL DESCRIPTION

This section, the first of two describing the mechanism, provides a general description of the entire mechanism. It includes a statement of the system's overall function, after which it names the main components in the order they will be used in the operation.

A drawing of the mechanism's major components accompanies this section, often on a foldout page that the reader can look at as he studies the manual. Exploded views of small or complex sub-parts are either inserted into the overall drawing or placed at appropriate places in the descriptive sections of the manual. On these illustrations, you give each part a reference number which you use throughout the manual and in the parts list.

DETAILED DESCRIPTION

The main components listed in the general description are broken down and described in the detailed description section. Use headings and subordination to clarify the relationships of the subparts to the major components. Rather than providing details about all the parts, concentrate on describing the physical relationships of particularly complex parts. You do this for two reasons: First, you can assume that the reader has the mechanism right in front of him; along with the mechanism itself, your overall and exploded visuals go a long way toward clarifying physical characteristics. Second, there is no need to provide information the reader does not require; the function of a manual's description sections is to give information the reader needs to understand subsequent sections on theory of operation, operation, and maintenance.

Some manuals combine the sections on general and detailed description. Other integrate detailed description with theory of operation because the components lend themselves to that type of organization. Such variations are justified if they take both the reader and the mechanism into consideration.

THEORY OF OPERATION

Contrary to what might be expected, principles of operation support the maintenance procedures more than the operating procedures. If the operating procedures are well-written, an operator can generally run the machine and handle emergency situations with limited knowledge of theory. With this in mind, some writers place the theory section immediately in front of the maintenance procedures rather than earlier in the manual.

Theory comes into play when an operator, having followed directions for checking a malfunction and still not found out what is wrong with the mechanism, says, "What do I do now?" He is forced to apply his knowledge of the mechanism's principles, or the logic of its operation, to solve the problem.

Recognizing the operator's background, try to present the theory of the mechanism in as nontechnical a way as possible. Rather than writing a textbook, explain the theory as it applies to the particular mechanism. In effect, the theory section zeroes in on the mechanism's particularly complex sequences, describes their operation, and integrates theoretical information into the description. If at all possible, mathematics should be avoided in favor of diagrammatic explanations of complex information. Illustrations, including diagrams with a limited number of symbols, help clarify the operational sequences' relationships to each other. You can refer to visuals in the description sections, but theory generally calls for a new set of drawings.

OPERATION

The type of writing used in a manual's operation and maintenance sections rarely appears elsewhere in technical writing. It consists of sets of directions which bluntly direct the operator to perform the necessary operations. Grammatically speaking, you abandon the active voice, indicative mood in favor of the active voice, imperative mood: "Place calibration tape on tape deck." The imperative mood implies the word *you* in front of each command. The sentences are very short and concise, and like the example above, somewhat abbreviated.

Just as headings and subordination are used in the preceding sections of the manual, lengthy operational sequences are broken into small, easy-to-follow sequences, as demonstrated in Section 5.0 of the model at the end of this chapter. To avoid repetition, you can sometimes tell your reader to duplicate a procedure. However, if the operator has to constantly flip back and forth in the manual, the space saved is not worth the potential confusion.

The unchanging sentence structure in the operation section of a manual, combined with your thorough knowledge of the mechanism, can cause almost automatic writing and lead to occasional omissions. Keep in mind that if you leave out a step, the operator is not likely to catch the error, and the result can be serious. For the same reason, place warnings about potentially dangerous operations in centered boxes for emphasis. They contain the word *warning* in capital letters and a short explanation of the danger.

Within the set of directions, references should be made to drawings in the description sections to help the reader locate controls and gauges. For complex procedures, visuals can be placed right in the set of directions.

MAINTENANCE

In self-contained manuals, you often integrate information about service and maintenance. You provide step-by-step procedures, called check-out procedures, for routine maintenance. At appropriate places within these procedures, refer to a troubleshooting chart which lists symptoms and possible solutions for malfunctions, as in the hypothetical example below:

PROBLEM	POSSIBLE CAUSE
Vacuum tube (7 - Fig. 2) fails to light	Volt circuit (Fig. 4) not connected. Broken connection in filament circuit (Fig. 5). See Testing Procedure 2.

In the troubleshooting chart, divide the problems according to major components, using the order of components established early in the manual. Identify each component by number, and refer the reader to the illustration in which it appears.

If the operator must use complex testing equipment to locate malfunctions, the maintenance section often includes directions for operating the equipment. Refer to these directions at appropriate places in the maintenance section, and reinforce intricate testing procedures with illustrations.

PARTS LIST

A parts list, or catalog, names and describes every component, no matter how minute, in the entire mechanism. The list helps the reader identify parts which need replacing and simplifies the pro-

cess for ordering them. The following hypothetical example shows
the list's categories.

Index Number	Manufacturer's Number	Description	Quantity
Resistors (Fig. 5)			
1	R 1-1	1 megohm, ½ watt	2
2	R 1-5	220 megohm, ½ watt	1
3	R 1-9	1,000 megohm, 1 watt	1

The index numbers are yours; you use them throughout the manual
to identify the parts. In the parts list, refer the reader to a visual
which shows the part, as the example's reference to Figure 5 indicates. Also provide the manufacturer's part number, a brief description of the part, and the number of parts in the mechanism.
By grouping the parts according to type (resistors, capacitors,
switches), and listing the individual components in the order of
their index numbers, you give your parts list a logical organization.

MANUAL
on the
T-1-A THEODOLITE SURVEYING INSTRUMENT

Prepared for
TORRENGA ENGINEERING

by
Richard K. Hardesty

May 13, 1972

TABLE OF CONTENTS

LIST OF ILLUSTRATIONS

1.0 INTRODUCTION

This manual provides operating instructions for Torrenga employees who are gaining practical surveying experience with the T-1-A theodolite. Operations are limited to those commonly performed in the field.

The manual's major sections are as follows:

General Description

Detailed Description

Theory of Operation

Operation

After acquiring experience with the theodolite, operators should examine the manufacturer's manual, Wild T-1-A[1], for more detailed and theoretical information.

2.0 GENERAL DESCRIPTION

The main function of the theodolite, a surveying instrument, is to measure horizontal and vertical angles. In the field, it is used for working from one point to another when shooting a line, and from one point to two other points when measuring an angle.

---

[1] The original manual is published by Heerbrugg Ltd., Heerbrugg, Switzerland. Surveying Practice, by Philip C. E. Kissam, published by McGraw-Hill, was also consulted during the writing of this manual.

In Section 3.0, the theodolite's components are grouped according to function and described in the order they are used during operation:

> Carrying Device
>
> Tripod
>
> Leveling and Centering Devices
>
> Components for Focusing
>
> Degree Circle
>
> Components for Turning Angles

3.0   DETAILED DESCRIPTION

3.1   CARRYING DEVICE

During transportation, the theodolite should be placed in a steel case (Fig. 1), which has a base plate (1) and a cover (2). Two clamps, one on each side of the base plate, keep the instrument stationary. The cover has two clamps, one on each side, which connect and hold the theodolite to the base plate. A leather strap, used for carrying the instrument from job to job, connects to the two clamps on the cover.

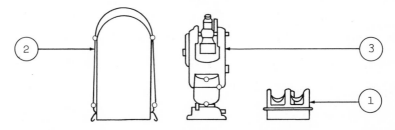

FIG. 1   THEODOLITE AND CARRYING DEVICE

3.2   TRIPOD

As shown in Figure 2, the theodolite (3) mounts on the head of a tripod (4). A screw underneath the tripod head holds the theodolite. Loosening the screw allows the instrument to move on the tripod head.

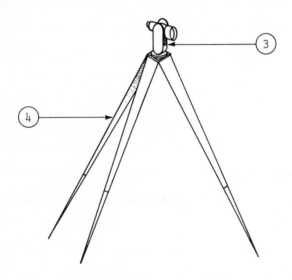

FIG. 2   THEODOLITE AND TRIPOD

3.3   LEVELING AND CENTERING DEVICES

To be accurate, the theodolite must have the same vertical axis as the point on which the tripod sits. To accomplish this, the operator must understand three components. (From here on, all components are shown in Figure 3, page 4.)

An optical plummet (5) appears directly below the instrument's scope. Through use of mirrors, this scope shows the point the tripod sits on. The operator must remember that everything seen through the scope appears to be opposite its actual direction. Things appearing to be too far to the left are really too far right.

On the instrument's base, a circular glass (6) engraved with a cross encases a liquid and an air bubble. To level the theodolite, footscrews (7) on the theodolite's feet are adjusted until the air bubble centers on the cross.

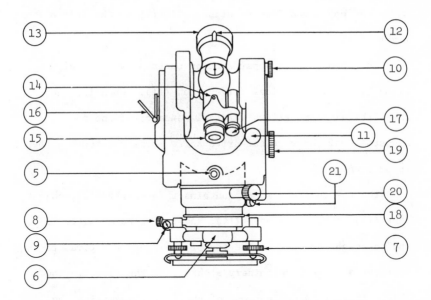

(5)	Optical Plummet	(13)	Object Lens
(6)	Circular Glass	(14)	Focusing Sleeve
(7)	Footscrews	(15)	Telescope Eyepiece
(8)	Lower-Plate Clamp	(16)	Illuminating Mirror
(9)	Lower-Plate Drive Screw	(17)	Microscope-Reading Eyepiece
(10)	Vertical Clamp	(18)	Milled Ring
(11)	Vertical-Drive Screw	(19)	Micrometer Screw
(12)	Foresight	(20)	Horizontal (Upper) Clamp
		(21)	Upper-Plate Drive Screw

FIG. 3   COMPONENTS OF THE THEODOLITE

3.4 COMPONENTS FOR FOCUSING

Focusing the theodolite requires understanding of eight components.

The theodolite's lower section contains a lower-plate clamp (8) and a lower-plate drive screw (9). When loosened, the lower-plate clamp allows free rotation of both the upper and lower sections of the instrument. When the lower-plate clamp is tightened, a turn of the lower-plate drive screw moves the upper and lower sections slightly to the left or right.

The theodolite's upper section contains a vertical clamp (10) and a vertical-drive screw (11). When loosened, the vertical clamp allows vertical movement of the scope. Adjustment of the vertical-drive screw moves the scope either up or down.

The theodolite's scope contains a foresight (12), object lens (13), focusing sleeve (14), and telescope eyepiece (15). The foresight appears on the top of the scope and gives the operator a rough, preliminary sighting of the target. The object lens, located in front of the scope, magnifies the target. The focusing sleeve, located at the center of the scope, is used for focusing the scope.

3.5 DEGREE CIRCLE

Figure 4 shows the instrument's three reading scales, called the degree circle, used for measuring angles.

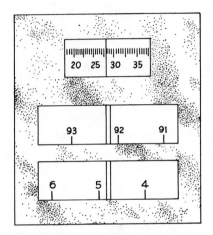

FIG. 4   DEGREE CIRCLE

The top scale shows the minutes and seconds of a degree.
The vertical-degree scale, for measuring vertical angles,
appears in the center of the figure; rarely used, it will
not be discussed in this manual. The horizontal-degree
scale, which is graduated into 360 degrees, appears at the
bottom of the figure.

3.6   COMPONENTS FOR TURNING ANGLES

The six components for turning angles are the illuminating mirror (16); microscope-reading eyepiece (17), located on the left-center of the theodolite; milled ring (18),
located between the upper and lower sections; micrometer
screw (19), located on the right side of the instrument
below the vertical-drive screw; and the horizontal (upper)
clamp (20), located on the instrument's upper section.

When adjusted, the illuminating mirror reflects light into the instrument to illuminate the degree circle. Section 4.0 explains the functions of the five other components for turning angles.

4.0 THEORY OF OPERATION

4.1 FOCUSING

To focus the scope's cross hairs on the target, the operator looks through the telescope eyepiece (15) and adjusts the focusing sleeve (14).

4.2 TURNING ANGLES

When loosened, the horizontal (upper) clamp (20) allows free movement of the upper section, and the upper-plate drive screw (21) allows very small and accurate movement of the upper section.

The operator looks through the microscope-reading eyepiece (17) to read the degree circle. The milled ring (18) allows the operator to advance or retard the circle for reading different-sized angles, and the micrometer screw (19) allows fine adjustment of the circle while measuring angles. On the degree circle (Fig. 4), the minute-second scale's vertical line and the horizontal-degree scale's double-vertical lines are used to find correct angles.

The degrees of an angle are read directly when the operator turns the instrument from zero degrees (clockwise); the

operator will learn variations of this procedure as he gains
experience. When the operator turns the instrument to zero
degrees (counterclockwise), he subtracts the angle he reads
from 360 degrees to find the correct angle.

5.0 OPERATION

5.1 SETTING UP THE INSTRUMENT

   1. Remove cover (2) from the theodolite (3).

   2. Remove the theodolite from the base plate (1).

   3. Place theodolite on the tripod (4).

   4. Insert screw on the tripod into the base of the theodolite and tighten.

5.2 LEVELING AND CENTERING THE INSTRUMENT

   1. Place the theodolite over the point where it will
be set up.

   2. Adjust the tripod's legs so that its head is somewhat level and positioned directly over the point.

   3. Adjust the footscrews (7) until the air bubble is
centered in the circular glass (6).

   4. Loosen the screw which connects the tripod to the
theodolite.

   5. Move the theodolite until it is directly over the
point.

   6. Tighten the screw which connects the theodolite to
the tripod.

5.3   FOCUSING

When focusing, remember to periodically check the level of the instrument.

1.   Loosen the lower-plate clamp (8).

2.   Rotate the instrument until the target appears in the foresight (12).

3.   Tighten the lower-plate clamp.

4.   Loosen the vertical clamp (10) and move the scope until the target is visible.

5.   Tighten the vertical clamp.

6.   Looking through the telescope eyepiece (15), adjust the focusing sleeve (14).

7.   Adjust the vertical-drive screw (11).

8.   Looking through the telescope eyepiece, adjust the lower-plate drive screw (9), putting the scope's vertical line exactly on the target.

5.4   TURNING AN ANGLE FROM ZERO DEGREES

1.   Adjust the illuminating mirror (16), putting light on the degree circle.

2.   Looking through the microscope eyepiece (17), adjust the micrometer screw (19), lining up zero minutes with the vertical line.

3.   Loosen the horizontal (upper) clamp (20) and the lower-plate clamp (8).

4. Rotate the milled ring (18) until zero degrees is as close as possible to the vertical line.

5. Tighten the horizontal (upper) clamp (20) and the lower-plate clamp (8).

6. Adjust the upper-plate drive screw (21) until zero degrees is exactly in line with the scope's vertical line.

7. Follow steps 1 through 8 of Instructions 5.3 above.

8. Loosen the horizontal (upper) clamp (20), and rotate the instrument until reaching the desired degree reading, and then tighten the horizontal clamp.

9. If a degree is not already lined up with the vertical line in the degree circle, adjust the micrometer screw (19) to line up the nearest degree.

10. Read the angle.

5.5 TURNING AN ANGLE TO ZERO DEGREES

1. Adjust the illuminating mirror (16), putting light on the mirror.

2. Looking through the microscope eyepiece (17), adjust the micrometer screw (19), lining up the desired minutes and seconds with the vertical line.

3. Adjust the upper-plate drive screw (2) until the degree of the angle being turned is lined up with the vertical line.

4. Follow steps 3 through 6 of Instructions 5.4 above.

5. Loosen the horizontal (upper) clamp (20) and rotate the instrument counterclockwise until zero degrees is as close as possible to the vertical line.

6. Tighten the horizontal (upper) clamp and adjust the upper-plate drive screw (21) until zero degrees is exactly lined up with the vertical line.

7. Adjust the micrometer screw (19), lining up zero minutes with the vertical line.

5.6 TAKING DOWN THE INSTRUMENT

1. Remove the screw which connects the theodolite to the tripod.

2. Take the theodolite off the tripod.

3. Clamp the theodolite to the base plate (1).

4. Clamp the cover (2) to the base plate.

EXERCISE

Write a step-by-step set of directions for changing a flat tire, aiming your writing at someone who has never done it. To make the directions easy to follow, break them into logical sequences such as (1) assembling jack, (2) positioning jack, (3) removing flat tire, (4) mounting new tire, (5) and disassembling jack. Your set of directions, which should total approximately 50 steps, may include the following two drawings.

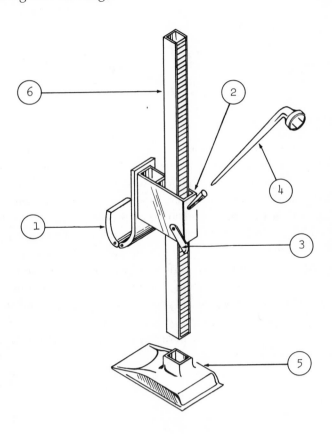

(1)	Jack Hook	(4)	Jack Handle
(2)	Jack Head	(5)	Jack Base
(3)	Jack Locking Lever	(6)	Jack Post

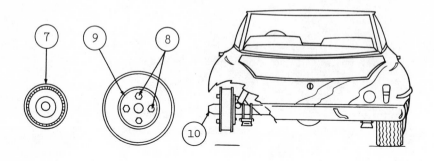

(7) Wheel Cover   (9) Rim

(8) Lug Nuts   (10) Hub

WRITING ASSIGNMENT

Select a laboratory instrument which you frequently use when working on projects in your major area. Write a 750-word manual aimed at a high school sophomore who has only basic technical knowledge and has not previously operated the mechanism.

Along with a title page and table of contents, include an introduction and sections on description (general and detailed), principles of operation, and operation. The operating instructions should be for procedures commonly performed with the instrument; identify the procedures in the scope section of your introduction.

Assume that the sophomore is your assistant and that the success of your next project depends on his operating skill. Carefully subordinate and number your paragraphs, and provide illustrations whenever necessary.

14. ORAL REPORTS

As you advance in industry, your ability to speak convincingly, as well as your writing skill, becomes increasingly important. At conferences, you are called upon to explain the results of investigations, propose solutions to problems, report on the progress of projects, and justify your department's requests for more men and equipment. Every type of formal, written report has its verbal counterpart; sometimes an oral report supplements a written one, and often a verbal presentation takes the place of written reports. Although you use the same information-gathering process for both oral and written communication, you must organize the information differently.

ORAL AND WRITTEN REPORTS

The most significant difference between oral and written communication is also the most obvious: an oral report has listeners, and a written one has readers. But why not use the same report for both types of presentations? How is a listener different from a reader?

1. A listener is present for the entire oral report. This would seem advantageous for a speaker, but it actually makes verbal communication more difficult. Let's assume that your listeners are the plant manager and his staff. It is unlikely that the plant manager wants to hear your entire report. He would probably prefer a capsule version of the report, and a written report's introductory summary would give him one. He would then order his staff members to examine the details contained in the body of the report. However, an oral report gives him no choice but to listen to all your information.

    The manager's staff members are not interested in the entire report, either. If the report were written, each of them would read the introductory summary and use the table of contents to locate the financial, technical, personnel, or other section of the report that interests him. Nevertheless, all of the staff members are present for the entire oral report.

2. Your listeners have only one opportunity to grasp the information. Even though a question-and-answer period may follow the

report, listeners cannot study the information as a reader would study a formal report.

3. An oral report does not provide headings to identify sections of particular interest to the listeners and to indicate parallel and subordinate ideas.

These differences emphasize the problems that exist in speaker-listener situations. However, a speaker also enjoys many advantages. He can use his personality, voice, and gestures, as well as first-person pronouns, visuals, and feedback from his listeners. The remainder of this chapter explains how you can effectively organize and communicate your information in an oral report.

THE EXTEMPORANEOUS REPORT

Many inexperienced speakers want to write every word of their reports. They are afraid of forgetting important ideas and think the solution is to either read or memorize the report. However, most experienced speakers agree that the extemporaneous report, which is outlined rather than read or memorized, is the most effective type of presentation. The extemporaneous method emphasizes natural, conversational delivery and concentration on the audience. Using this method, you can direct your attention to the listeners, referring to the outline only to jog your memory and insure that ideas are presented in the proper order. Your word choice occasionally suffers, but words spoken do not necessarily equal ideas communicated, anyway.

The best way to combat fear and forgetfulness during an extemporaneous report is simply to know your subject. If you are the best-informed person in the room, you can stop worrying about your subject and start concentrating on communicating. The outline for an extemporaneous speech contains main headings and subheadings; development of the headings into sentences is accomplished during rehearsals.

THE INTRODUCTION

In industry, do not worry about finding a humorous story, a quotation by an authority, or a recent occurrence to begin your report. At the outset, you already have your listeners' attention. Capitalize on that by saying something strongly related to your topic. If your opening comments are interesting or amusing but not relevant to the report, you have to add a second attention step to transfer the listeners' attention to your subject. A good opening may simply identify your subject and purpose, show how the subject affects the listeners, or explain why the investigation was undertaken.

The introduction must also give your listeners a clear idea of the order in which you intend to present your ideas. For a recommendation report, this requires stating the possible alternatives and specifying your criteria for judging them. You can introduce a proposal by identifying the major areas that you will discuss. In any oral report, the listeners must clearly understand your plan of development so that they can grasp the information in the body of the report.

THE BODY

Time is to an oral report what space is to a written one. However, an oral report does not lend itself to the concise type of presentation you use in a written report. To verbally communicate an idea, you must state a generalization, provide details to support it, and reinforce it with a summary. Numerous studies have shown that listeners simply do not hear everything you say, and if they miss an idea or an important detail, they have no recourse. Therefore, the verbal communication process consumes a minimum of five minutes per main idea.

Always impose your own time limit on the report, and narrow your number of main ideas accordingly. It is much better to carefully present two or three main ideas than to attempt to communicate more information than your listeners can grasp. If you select only the most important ideas, your speech will be limited enough to please the plant manager and detailed enough to satisfy his staff members. Throughout the report, avoid potentially confusing terms, and if possible, reinforce your main ideas with visuals.

CONCLUSIONS

As you conclude your report, actually say, "In conclusion . . ." to capture your listeners' interest. The concluding section emphasizes the main ideas presented in the body of the report. If your objective is persuasion, stress the ideas' main advantages and urge your listeners to take specific action. For a recommendation report, emphasize the most significant data presented for each criterion and clearly present your recommendations. You can often use a visual to summarize the important data presented in either a proposal or a recommendation report. End the report by asking if your listeners have any questions.

VISUALS

Visuals are invaluable for emphasizing and clarifying complex ideas in an oral report. As you construct your outline, determine which ideas will benefit from visuals and build the report with them in mind. The visuals should be aimed at the technical level of your listeners; as Chapter 8 explains, the same information can be presented in either simple or complex form. Also, use several simple visuals instead of crowding too much information on one, and place as few words as possible on the visuals—you are responsible for verbal interpretation. Color adds life to visuals, but black and white offers clarity if some listeners are far away.

The size of the audience partially determines the method of showing or projecting visuals. Flip-charts are easy to manipulate and effective for small groups. For small or large groups, consider slide, opaque, or overhead projectors. The size of the projected images can be regulated by simply moving the projector toward or away from the screen and adjusting the focus. The room must be semi-dark for slide and opaque projections. However, overhead projections can be used in normally-lighted rooms; nobody has to flip light switches, and the audience is not distracted. Overhead projectors require transparencies, but making them is a simple process, and they are easy to handle during the presentation.

REHEARSALS

To be successful, extemporaneous reports must be thoroughly rehearsed. During rehearsals, go straight through the speech, using small note cards. In each trial run, attempt to give the presentation a conversational quality, and practice using your voice and gestures to emphasize important points. You have rehearsed enough when you feel secure with your report, but always stop short of memorization; if you do not, you will ultimately grope for memorized words rather than concentrate on the listeners and let the words flow.

Final rehearsals for reports to large groups should simulate conditions under which the speech will be made. This includes a room of approximately the same size with the same type of equipment for projecting your voice and visuals. Rehearsals of this type not only guard against technical problems but allow you to become comfortable in an environment similar to the one for the final report. Ask a colleague to attend at least one rehearsal, comment on how well he can hear you and see the visuals, and offer a critique of the speech, including any of your distracting mannerisms.

DELIVERING THE REPORT

A report can be no better than its preparation, but the following is a list of suggestions to keep in mind as you face your listeners and deliver the speech.

1. Do not start the speech with an apology. An audience is usually more willing to accept your information if you speak positively.
2. Make sure you can be heard, but try to speak conversationally. The listeners should get the impression that "he is talking to me" rather than that "he is making a report." Inexperienced speakers often talk too rapidly.
3. Look directly at each listener at least once during the report. With experience, you will be able to tell by the viewers' faces whether you are communicating.
4. Fight the tendency to use your outline when you do not need it. When collecting your thoughts, do not say "uh"; pause and remain silent.
5. When finished with a visual, remove it so that it does not compete with you. If you are using a pointer, set it down to avoid tapping with it.
6. If you cannot answer a question during the question-and-answer session, say so and assure the questioner that you will find the answer for him.

SPEAKING ASSIGNMENT

Your instructor may require an oral presentation of a formal report you have written during the term. The speech should be extemporaneous and approximately 10 minutes in length. To prepare your presentation, follow the suggestions in this chapter for converting a written report into a successful oral presentation. Make your visuals, outline the speech, and most important, rehearse. A question-and-answer session with other members of the class should follow your presentation.

SECTION FIVE

LETTERS

LETTER FORMATS / TYPES OF BUSINESS LETTERS / APPLICATION LETTERS

SECTION FIVE OVERVIEW

In this section, the text moves away from essentially data-oriented reports to people-oriented letters. First, the various letter formats and the elements of a letter are described. Then, the types of business letters are explained, with emphasis on the indirect approach for letters of persuasion, and the direct approach for informative letters. Application letters, perhaps the most important letters you will write, are emphasized in the final chapter.

15. LETTER FORMATS

A letter's appearance gives the reader his first impression of the writer and the firm he represents. Contributing to the visual attractiveness of a letter are its letterhead, balance, neatness, and format. The format must also be appropriate to the letter's contents. This chapter begins by discussing the common formats for business letters: the block, full-block, modified-block, and simplified formats. Then, the elements that comprise a letter's format are described and exemplified.

THE FOUR FORMATS

In the block, full-block, and modified-block formats, the word *block* refers to the shape of the letter's major sections. When the left side of a section has no indentation, the section is considered block-shaped.

The *block format's* inside address, body, and signature sections have no indentation (Fig. 1, p. 264). The block for the inside address and body are flush with the left margin, and the left side of the signature block begins at the center of the page. As with the full- and modified-block formats, if the letter contains no letterhead, its heading includes the writer's address as well as the date, and they form a block which ends at approximately the right margin.

In the *full-block format* (Fig. 2, p. 265), each element of the letter is flush against the left margin. Thus, the left side of the letter forms a full, or complete, block. The full-block format lacks balance but is probably the most widely used because it can be quickly typed.

Placement of the blocks in the *modified-block format* (Fig. 3, p. 266) is identical to that of the block format, which causes some confusion between the two. However, the first line of each paragraph in a modified-block format is indented, modifying the block. Another name for this format is the "semiblock format."

The *simplified format* (Fig. 4, p. 267) is a recent development which departs from conventional letter formats. Its streamlined format contains no salutation and no complimentary close, but almost always includes a subject line. Advocates of this format consider it more efficient than the other three because it requires

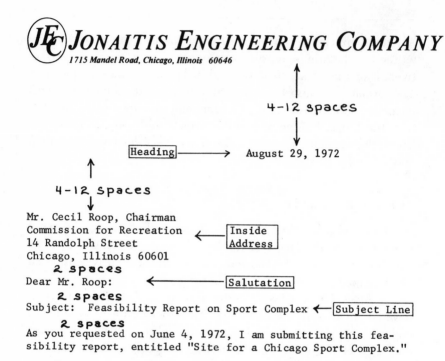

JONAITIS ENGINEERING COMPANY
1715 Mandel Road, Chicago, Illinois 60646

4-12 spaces

Heading ⟶ August 29, 1972

4-12 spaces

Mr. Cecil Roop, Chairman
Commission for Recreation   ⟵ Inside Address
14 Randolph Street
Chicago, Illinois 60601

2 spaces

Dear Mr. Roop: ⟵ Salutation

2 spaces

Subject: Feasibility Report on Sport Complex ⟵ Subject Line

2 spaces

As you requested on June 4, 1972, I am submitting this feasibility report, entitled "Site for a Chicago Sport Complex."

The report examines the feasibility of three proposed sites:
Solder Field, the Near West Side, and the South Loop. As
you directed, emphasis has been placed on traffic accessibility and pre-construction costs, and the recommendations
are based on those criteria. New city expressways are
presently being proposed, but they remain an uncertain variable and are not considered in this study.

I am available for consultation about this report and for
further research as you move forward with the sport complex
project.

2 spaces

Complimentary close ⟶ Sincerely yours,

4 spaces

Signature ⟶ Robert W. Jonaitis

Fig. 1 *Block Format*

JONAITIS ENGINEERING COMPANY
1715 Mandel Road, Chicago, Illinois 60646

August 29, 1972

Mr. Cecil Roop, Chairman
Commission for Recreation
14 Randolph Street
Chicago, Illinois 60601

Dear Mr. Roop:

Subject: Feasibility Report on Sport Complex

As you requested on June 4, 1972, I am submitting this feasibility report, entitled "Site for a Chicago Sport Complex."

The report examines . . . .
. . . .

Sincerely yours,

Robert W. Jonaitis

Fig. 2 *Full Block Format*

JONAITIS ENGINEERING COMPANY
1715 Mandel Road, Chicago, Illinois 60646

August 29, 1972

Mr. Cecil Roop, Chairman
Commission for Recreation
14 Randolph Street
Chicago, Illinois 60601

Dear Mr. Roop:

    Subject: Feasibility Report on Sport Complex

    As you requested on June 4, 1972, I am submitting this
feasibility report, entitled "Site for a Chicago Sport Complex."

    The report examines . . . .
. . . .

                     Sincerely yours,

                     Robert W. Jonaitis

Fig. 3 *Modified Block Format*

JONAITIS ENGINEERING COMPANY
1715 Mandel Road, Chicago, Illinois 60646

August 29, 1972

Mr. Cecil Roop, Chairman
Commission for Recreation
14 Randolph Street
Chicago, Illinois 60601

Subject: Feasibility Report on Sport Complex

As you requested on June 4, 1972, I am submitting this feasibility report, entitled "Site for a Chicago Sport Complex."

The report examines the feasibility of three proposed sites:
Soldier Field, the Near West Side, and the South Loop. As
you directed, emphasis has been placed on traffic accessibility and pre-construction costs, and the recommendations
are based on those criteria. New city expressways are presently being proposed, but they remain an uncertain variable
and are not considered in this study.

I am available for consultation about this report and for
further research as you move forward with the sport complex
project.

Robert W. Jonaitis

Fig. 4 *Simplified Format*

less typing. However, a letter's efficiency should be determined by its results rather than its typing time. The success of a simplified letter depends on the writer's ability to overcome the format's extremely direct approach and establish rapport with his reader. Rapport is not crucial in letters containing routine or positive information, but the simplified format would seem to present problems in letters requiring persuasion. Such letters need an indirect approach, and there is nothing indirect about a subject line. Direct and indirect approaches will be discussed more specifically in Chapter 16.

ELEMENTS OF A LETTER

As Figure 1 shows, letters of any length can be given vertical balance by adjusting the space between the letterhead and the date, and between the date and the inside address. Horizontal balance is achieved by widening or narrowing the letter's margins. The body of a letter should be single spaced, with two spaces between paragraphs. The parts which make up a business letter's format are described below.

HEADING

Because virtually all companies use letterhead stationery, the heading of a business letter contains only the date, which ends at about the right margin. In personal letters, include both the address and the date:

```
4217 East Avenue
Claremont, Indiana 46327
May 15, 1972
```

A complete personal letter appears at the end of Chapter 17.

INSIDE ADDRESS

Readers are sensitive about their names, titles, and firms, so the inside address requires special care. Make sure that you use the correct personal title (Mr., Dr., Professor) and business title (Director, Chairman, Treasurer). Write the firm's name exactly as they do, adhering to their practice of abbreviating or spelling out such words as "Company" and "Corporation."

The reader's business title can be placed after his name, on a line by itself, or preceding the name of the firm; select the placement that best balances the inside address.

```
Mr. Byron Collins
Senior Engineer
American Rail Company
9115 Charles Avenue
Lee, California 93541
```

ATTENTION LINE

Attention lines, which appear two spaces below the inside address, are generally used only when you cannot name the reader (Attention Personnel Manager; Attention Payroll Department). Today, the word "Attention" is usually placed against the left margin, and is not followed by a colon.

SALUTATION

The salutation always agrees with the first line of the inside address. If the first line contains the name of a company, your salutation must be plural (Gentlemen:), regardless of the attention line. If the inside address names an individual (Mr. Byron Collins), say "Dear Mr. Collins" or "Dear Sir." A colon always follows a business letter's salutation.

```
American Rail Company
9115 Charles Avenue
Lee, California 93541

Attention Mr. Byron Collins

Gentlemen:
```

SUBJECT LINE

Subject lines, which allow a direct approach to the subject of the letter, are becoming common in business letters, particularly information letters. Place the subject line (Subject: Training Program) two spaces below the salutation, and indent it if you indent the letter's paragraphs. Subject lines are often either underlined or completely capitalized.

COMPLIMENTARY CLOSE AND SIGNATURE

Use simple closings, such as "Sincerely yours" or "Yours truly," to end business letters. Capitalize only the first word, and place a comma after the closing.

Company policy generally specifies what to place in the signature section. It sometimes requires that the company's name be placed immediately below the complimentary close (see below left). The writer's title or department, or both, often appear below his typed name (see below right).

Yours sincerely,
DAVIS MANUFACTURING CO.

Ronald C. Anderson

Ronald C. Anderson
Personnel Director

Yours truly,

William J. Brockington

William J. Brockington
Manager, Drafting Division

SUCCEEDING PAGES

For succeeding pages of a letter, place the name of the reader, the page number, and the date in a heading:

Mr. Collins   -2-   May 15, 1972

16. TYPES OF BUSINESS LETTERS

Awareness of the reader, which has been emphasized in each of the report-writing chapters, is even more important in business letters. Although letters and reports both require anticipation of the reader's needs, letters are generally more people-oriented: the letter reader's reaction to the *way* you present information is often more important than the information itself. In letters and in everyday conversation, how many times have you been disturbed not so much by what someone has said, but by the way he said it? This chapter explains how to word and organize your letters so they will achieve their purpose and avoid alienating the reader.

STYLE

In business letters, use the first person "I" and "we," and refer to the reader in the second person "you." This is the first step toward giving your letters the natural, conversational style needed for personal communication. The second step is to write in plain English. The "businessese" style of writing convinces your reader that you regard him as a number rather than as an individual:

> Pursuant to our discussion of February 3 in reference to the L-19 transistor, please be advised that we are not presently in receipt of the above-mentioned item. Enclosed herewith please find a brochure regarding said transistor as per your request.

Such stilted, pompous prose is so common in business correspondence that many young writers think they are supposed to write that way. Why not say, "I've enclosed a brochure on the L-19 transistor we talked about on February 3. Our shipment of L-19's should arrive within a week." Recall the last bit of correspondence you received from your school; did it sound like a human being had written it? When you write letters in business and industry, remember that a conversational style invariably achieves better results than "businessese."

TONE

Tone refers to your attitude toward the subject and reader, as re-
flected in your choice of words and your overall approach to the
letter-writing situation. When you write a letter, you are the mes-
sage; you must demonstrate your understanding of the reader's
point of view by using the "you approach," and you must also use
a "positive approach" to reveal your company's point of view.

THE "YOU APPROACH"

On one level, the "you approach," requires only a common-sense
awareness of human nature. No reader wants to be treated like a
number, taken for granted, told how to do his job, or threatened.
On the other hand, human nature occasionally tempts all of us to
do those very things, particularly in frustrating letter-writing situations. When that happens, remember that your letter represents not
only yourself but your firm. The "I-guess-I-told-him" letter that
sometimes makes you feel good generally does not get the results
your company wants. The reader will be alienated unless you show
him common courtesy and respect.

On another level, the "you approach" requires a thorough analysis of each reader's point of view. His point of view is invariably
different from that of your firm. For example, let's assume that your
firm has sold a conveyor system to Calvin Number. To your firm,
Number represents just another conveyor system, a very small item
on your profit-and-loss ledger. For Number's company, of course,
the conveyor might represent last month's profits. If the conveyor is
defective, it becomes one of the many routine problems that your
firm handles every day. However, the defective conveyor might not
be routine for Number; perhaps he is very worried about it. To
maintain a good business relationship with him, your letter must
show your understanding of his position. The defect matters to
him, and he wants to know that it has some importance for you,
also. By putting yourself in Calvin Number's position, you will be
able to anticipate his reaction to your letter and to insure that you
do not evoke a negative response. A "you-approach" letter requires
time and thought, but it helps establish and maintain good business relationships.

THE "POSITIVE APPROACH"

Industry has no room for negative thinking, and it has even less room for negative letter-writing. From your firm's point of view, letters to customers and other firms are part of public relations; ultimately, they can affect profits. Therefore, all business letters are sales letters which promote or protect the image of your firm. To this extent, they are similar to interfirm proposals, which attempt to win contracts by communicating a "we-can-handle-it" attitude. In letters which must convey negative information about your firm or one of its customers, never misinform the reader but present the information in a positive rather than negative light. The following phrases should be avoided in business letters because they have negative connotations:

> This problem . . . (admits that there *is* a problem)
> You won't be sorry . . . (suggests that he *might* be sorry)
> Your failure . . . (implies that he is a failure)
> I doubt . . . (conveys negative attitude; say "I believe . . .")
> You claim . . . (implies that he has lied)

If you are not convinced of industry's emphasis on the "positive approach," try beginning an application letter with "I'm afraid I haven't had much actual working experience . . . ," and see what happens, or more specifically, see what does not happen.

DIRECT OR INDIRECT STRUCTURE?

The main function of every business letter is either to inform or to persuade. To determine the main function of your letter, decide whether the reader will react positively or negatively to the information it contains. If he will react negatively, give the letter an *indirect structure:* spend at least a paragraph preparing him for the information or persuading him to accept the information. On the other hand, if he will respond positively, give the letter a *direct structure:* present the information immediately. The form, or structure, of a letter is always determined by its main function; form follows function.

The rest of this chapter explains the functions of the most common types of business letters and suggests how their contents should be structured.

LETTERS OF INQUIRY

Firm-to-firm letters of inquiry, whether they are solicited or unsolicited, should be given a direct structure. Usually, you are asking about another firm's product or service, and you do not have to convince the reader to provide the information; firms view letters of inquiry as opportunities to gain customers. Therefore, simply identify the information you need in a one- or two-sentence letter.

As a student, you may occasionally write a personal letter of inquiry. Although such letters usually request information or some other favor, you can still use the direct approach. Large firms receive many requests of this type and appreciate a courteous but concise letter. State specifically the information you need, why you need it, and why you selected the particular firm as a source of information. In closing the letter, you may say that you look forward to hearing from the firm, but do not thank them in advance. Also, do not enclose a stamped, self-addressed envelope unless you are writing to an individual or to a small firm. Figure 1 illustrates an unsolicited, firm-to-firm letter of inquiry which requests special information.

ANSWERS TO INQUIRIES

Except for a letter responding to a student's request, an answer to an inquiry is usually a sales letter. Unlike most sales letters, however, it does not require an indirect approach; having solicited information, your reader will be receptive to it. Answer an inquiry immediately, if only to acknowledge the letter and explain that you have referred it to another department or dealer. An inquiry gives you some insight into the reader's needs, so inject the "you approach" immediately after thanking him for his letter. Describe your product in terms of his advantage, emphasizing its strong points. Enclose a brochure or other sales literature with your letter, and insure that he will have no difficulty ordering the product.

Larson
Electronics
Corp.

190 Jackson Blvd.
Atlanta, Ga.   30307
404-627-6335

April 5, 1972

Mr. Lawrence M. Callison
Personnel Director
Carlton Bridge Company
9311 Commerce Avenue
Boston, Massachusetts 02107

Dear Mr. Callison:

May I ask a favor of you? James Creekmore, a representative of the Melton Corporation, recently told me of your success in managing Carlton's training program for the hard-core unemployed. Our firm is presently planning such a program, and we would appreciate information about the attitude-changing section of your program.

We expect to begin our training sessions on approximately June 1. I believe we are prepared for the remedial-education and job-skills portion of the program, but we need help with attitude-changing. Specifically, we would appreciate knowing who you employed to teach that section, the major emphasis in the section, and the length of the section in relation to the entire program.

I congratulate you on your success in managing Carlton's program, and I believe your ideas would help us solve this long-neglected problem, also. I look forward to hearing from you.

Sincerely yours,

James F. Weathers

James F. Weathers
Project Coordinator

Fig. 1 *Letter of Inquiry*

TRANSMITTAL LETTERS

A transmittal letter conveys a report from one firm to another. It is an information letter which has a direct structure. Begin the letter by identifying the enclosed report and stating the date it was requested. Then, provide a brief paragraph or two explaining the report's purpose and scope. Close the letter by indicating your availability if the reader has any questions about the report. Chapter 9 describes transmittal correspondence in detail, and Chapter 15 contains an example of a transmittal letter.

CREDIT LETTERS

A credit letter responds to a request for credit from a firm or individual. The structure of the letter depends on whether your response is positive or negative. A letter *granting* credit should be given a direct approach; the reader will obviously be receptive to your information. Welcome him as a credit customer and explain the terms of credit. When *refusing* credit, however, an indirect structure is necessary. Thank him for requesting credit, and avoid insulting phrases like "bad risk" or "inability to pay debts promptly." In the second or third paragraph of your letter, inform him that you are unable to grant his request, but use the positive approach. Encourage him to buy on a cash basis until he has built up his credit rating, and express the hope that you will soon be able to extend credit to him.

ADJUSTMENT LETTERS

An adjustment letter responds to a customer's claim that your firm owes him something. The customer generally believes that you have sold him a defective product, and he wants you to replace it, repair it, or return his money. If your investigation shows that his claim is justified, thank him for calling it to your attention, apologize for the mistake, and tell him the terms of the adjustment. Never guarantee that the mistake will not happen again unless you are prepared to honor the guarantee.

A letter refusing adjustment must have an indirect structure. Thank the reader for his letter, express your understanding of his situation, and explain what action you have taken on his behalf.

Gradually build up to your refusal by explaining why the adjustment cannot be granted. Close the letter by stating that your firm appreciates his patronage, and looks forward to future business with him.

SALES LETTERS

Except in response to an inquiry, a sales letter must have an indirect structure. Readers receive hundreds of sales letters and must be enticed into reading them. A sales letter contains the following elements:

Attention: arouses the reader's interest in the product

Desire: uses the "you approach" by describing the product in terms of the reader's advantage

Conviction: provides details to convince him that the product is the best of its kind

Action: urges him to buy the product and makes it easy for him to do so

Apply the "you approach" throughout the letter but particularly in the "Attention" and "Desire" steps. In firm-to-firm sales letters, factual appeals to the reader's desire for profit, economy, and efficiency are the most effective. Firm-to-individual letters, in addition to emphasizing economy, generally appeal to the reader's desire for such things as prestige, sexual attractiveness, and comfort.

COLLECTION LETTERS

Depending on how a company views its customers, it uses one of two methods for collecting money. It either makes a strong effort to immediately collect the money, or it attempts to keep the customer *and* collect the money. The direct approach simply threatens the reader with either a suit or a collection agency. The indirect approach tries to avoid demanding the money. A firm which uses this approach wants the reader to do three things: keep the product, pay for it, and buy more products from the firm. The company's first collection letter assumes that the reader has forgotten the payment or wants to be reminded of it. The next letter assumes that the customer has not paid because the product is defective. Finally, as a last resort, the firm threatens the reader. When the indirect approach works, it generally accomplishes the firm's three goals. On the other hand, a direct letter often loses the customer even when it succeeds.

WRITING ASSIGNMENTS

1. You are a research analyst at Gabriel Corporation, which is building a carbon monoxide plant near Boise, Idaho. The Occupational Safety and Health Act of 1970 requires that the new plant have a safety program for checking employees' exposure to carbon monoxide. At the Environmental Research Center in Cleveland, Ohio, Dr. Marlon T. Jenner has been conducting tests on the effects of carbon monoxide. The results of his studies have been published in several periodicals. Write a letter to Dr. Jenner requesting that he assist you in setting up the safety program. Garbiel Corporation is prepared to pay Jenner's fee for consulting.

> Dr. Marlon T. Jenner
> Director of Research
> Environmental Research Center
> 3300 West 19th St.
> Cleveland, Ohio 44109

2. Mel Grant, a lift truck operator for Metro Compressed Gas Company, has been fired for working while under the influence of alcohol. Because Grant was extremely popular with the other employees, his dismissal has damaged their morale. Unfounded rumors that other firings are imminent have made the problem worse. In your position as department supervisor, you first considered using the regular company bulletin to inform employees of the facts. However, you rejected this idea because the bulletin would probably be viewed as company propaganda.

   Write a letter to William Swikert, the informal leader of the employees, explaining Grant's firing to him and requesting that he present the company's view to the workers.

Mr. William Swikert
2367 Livingston Ave.
Chicago, Illinois 60618

3. Arnold Engineering, where you are Assistant Manager, has been battling Electronic Systems for contracts to produce magnets. Several months ago, both firms bid for the job of supplying Auto Enterprises with a large quantity of motor magnets. Arnold won the contract and immediately ordered five tape applicators necessary for producing the magnets. Yesterday, your firm was informed that the applicators will arrive eight weeks late. Arnold Engineering's reputation for meeting deadlines is in jeopardy and the only tape applicators in the region are owned by Electronic Systems.

    Write a letter to Roger Duncan, general manager of Electronic Systems, convincing him to lease five tape applicators to your firm.

Mr. Roger Duncan
General Manager
Electronic Systems
4022 NW Eighth St.
Oklahoma City, Oklahoma 73107

4. Your former assistant at Data, Inc., was hired as personnel manager of Southwest Data Processing six months ago. In the past three months, five experienced engineers have quit your firm and gone to work for Southwest. It is now rumored that five more engineers have been contacted, and you fear that they may also switch to the competing firm. As personnel manager of Data, Inc., you have been asked by the vice president to write letters to five of your top engineers in an attempt to stop the employee thievery.

    Write a letter to John Ramsey, a bright young engineer who has been employed by Data, Inc., for four years.

Mr. John Ramsey
3991 Wilmore Blvd.
Springfield, New Mexico 87747

5. Sierra Ski Company has recently learned why its sales in the Colorado region fell so sharply during the past season. Sierra's Colorado distributor employed several salesmen who spent more time on the slopes than in the ski shops serving their accounts. As sales manager for Sierra, you have personally taken over the accounts of the best Colorado customers. You have decided to offer a 10 percent discount on their first orders for the new season. Write a sales letter to Gary Neal, an old customer whose purchases of Sierra equipment dropped almost 50 percent last season.

> Mr. Gary Neal
> Neal Ski Shop
> 3963 Lanley St.
> Denver 11, Colorado 80210

6. A year ago, McLaren Laboratories purchased five temperature controllers from Murcer, Inc. McLaren installed four of them and kept one as a spare. Yesterday the spare was installed and found to have a defective magnetic amplifier. As assistant manager, you called a Murcer salesman and asked for a replacement. Your request was refused because the unit's warranty has expired. Write a letter to William Carr explaining why you think the amplifier should be replaced.

> Mr. William Carr
> General Manager
> Murcer, Inc.
> 4616 Hawthorne Avenue
> Spokane, Washington 99205

7. When the supervisor of purchasing at Lombard's retired, James Linwood expected to be promoted. He had served very effectively as assistant supervisor for three years. However, the job was given to a brother-in-law of one of the company directors. Linwood immediately resigned and took a better position with a firm across town. The brother-in-law's incompetence soon became obvious, and he is now being trained to replace the aging Doberman pinscher that guards the company helicopter.

As personnel manager, you must write to Linwood, persuading him to return to Lombard's as supervisor of purchasing.

Mr. James Linwood
1455 Birch Drive
Old Orchard Beach, Maine 04064

17. APPLICATION LETTERS

A successful application letter must be preceded by self-analysis and an investigation of prospective firms. Both are prerequisites to the "you approach" explained in Chapter 16. In analyzing yourself, try to adopt the objectivity of prospective employers and determine what you have that will appeal to their self-interest. As a student, your strongest selling point is your education; therefore, you will devote most of the letter to describing it in terms of the potential employer's advantage. However, you should consider mentioning additional facts which might make you more valuable than other applicants. Employers generally prefer a student who has working experience related to his technical field. They also find it advantageous to hire someone who has a favorable draft status, has earned his college expenses, or is married. Such things as the ability to work with others or leadership potential should not be mentioned; the reader, who is experienced in evaluating applications, will judge them himself after examining the letter and data sheet.

After deciding what skills to emphasize, you can promote them much more effectively if you are aware of their market. The "you approach" cannot be used to its full potential unless you know how your skills might benefit each employer. A letter that demonstrates knowledge about a firm immediately distinguishes you from your competition; most applicants know only that a job is available. Through investigation, you can also find out which firms offer opportunities for advancement and reject the ones that don't.

There are many sources of information about firms: school placement offices, where company literature and the *College Placement Annual* are available; career booklets such as *Careers in Technology;* professional journals; and professors. The investigative process should begin long before application letters are written. After years of preparing for employment, no student should sell himself short by failing to examine the market thoroughly.

ORGANIZATION OF THE LETTER

The application letter contains three or four paragraphs, making it long enough to emphasize your main qualifications but short enough to invite reading. Its only function is to get an interview, at which you can elaborate upon your qualifications. It fulfills this objective by gaining the reader's attention, stating your main qualifications, and asking the reader for action—an interview.

ATTENTION

Although an application letter is essentially a sales letter, the indirect approach is not necessary. You can safely assume that your reader will be receptive to a brief, straightforward letter. Therefore, begin with a direct, positive paragraph which immediately specifies the job you are pursuing and summarizes your main qualifications. If you are not sure that an opening exists, apply for a particular type of work that the firm is engaged in. The "name beginning" provides a convenient opening sentence for the letter. It identifies the person, perhaps a school placement officer or a business associate of the firm, who suggested that an application letter be written. For the "name beginning" to work, the reader must be familiar with the name, and you must have received permission to use it. If this approach is impossible, start by simply asking that your qualifications for a certain job be considered. The first paragraph concludes with a one- or two-sentence summary of the particular skill you have to offer. As you emphasize your primary selling point, inject the "you approach," which is exemplified in the following samples of opening paragraphs:

> Professor J. V. Elton, head of the Department of Mechanical Engineering at Parker University, has informed me of your opening in the area of fluid dynamics. I have specialized in fluid dynamics at Parker and am eager to learn more about this opportunity and your company.

> I would like you to consider my qualifications for the position of corrosion engineer on your gas storage wells. I believe my three years' experience in this field can be of value to you.

> Has your Sterling Oaks Project opened up any summer jobs in your drafting department? If so, I am eager to show you what I can do, and you can help me determine what courses would be beneficial during my junior year in Parker University's School of Engineering.

AMPLIFICATION

The second section of an application letter, which takes up one or two paragraphs, offers concrete support for your initial statements. If you have emphasized your education, elaborate on it, perhaps by naming specific courses or projects that relate to the position you are seeking. This section is the longest one in the letter, but only the most pertinent facts merit inclusion. Always refer the reader to your attached personal data sheet for additional information or for support of a secondary selling point. In the examples below, the first writer emphasizes his education, and the second elaborates on his experience:

> I will receive my Bachelor of Science Degree in Mechanical Engineering this June. In the area of Fluid Dynamics, I have taken such courses as Gas Dynamics, Thermodynamics, Fluid Dynamics, and Fluid Mechanics. Other courses I have had which would prove helpful in this position are Instrumentation and Machine Design. The enclosed resume lists my working experience in the field of mechanical engineering.

> For the past three years, I have been employed as a corrosion engineer by Western Public Service Company. I have been in charge of the company's corrosion-prevention program at the Manchester gas-storage field for two years, and have worked on other piping systems. I am currently attending Parker University, taking evening courses toward a Bachelor of Science degree. The attached data sheet provides additional details about my experience and education.

ACTION

The function of an application letter is not to secure a job, but to get an interview at which the job and your qualifications can be discussed in detail. Therefore, your final paragraph asks for an interview at the reader's convenience. If there are periods when you will not be available, give the reader this information.

Many firms are reluctant to invite a student to come 500 or 1,000 miles for an initial interview. To get an interview with a distant firm, you can either suggest a meeting with a local representative, or mention that you will be visiting the firm's area during a certain period, perhaps a school vacation, and would appreciate an interview then. The firm's response will indicate whether traveling expenses will be paid, but you should be prepared to pay your own expenses if you suggest this type of arrangement. The following are examples of closing paragraphs:

> I would like to discuss my qualifications with you at a personal interview. I can be reached at WEstmore 4-4063.

> I will be visitng the Akron area during spring vacation, March 5-14, and would appreciate having an interview with you. I can be reached at the address or telephone number listed on my personal record.

> May I show you samples of my work? I am available for an interview at your convenience.

PERSONAL DATA SHEET

The data sheet complements a concise letter by providing a detailed history of the applicant. In contrast to the letter, which is aimed at individual firms whenever possible, the data sheet usually remains the same; Xeroxed copies are permissible if numerous applications are being sent. Readers prefer data sheets made up of concise, visually attractive data which allow quick assimilation. Needless to say, the data sheet must be complete. A year unaccounted for gives a negative impression.

As the example at the end of the chapter shows, the top of the data sheet contains your name, address, and telephone number, along with the date. If you include a photograph, place it in the upper-left corner. The rest of the data fall under four headings: "Personal Details," "Education," "Experience," and "References."

Items in the "Personal Details" section include age, height, weight, health, marital status, draft status after graduation, hobbies, and if applicable, church affiliation. Additional items may be placed in this section if they do not seem to fit under the other headings. For example, college organizations, honors, and activities belong here if you restrict the "Education" section to academic accomplishments.

The second heading is "Education" unless you have worked for several years and want to emphasize your experience. Educational details should be listed in reverse chronology to accentuate the most recent achievements. State the name of the school and degree, along with your date of graduation and grade-point average. Then, list the advanced courses which especially support the qualifications emphasized in your application letter.

The "Experience" section is also arranged in reverse chronology. It includes dates, names and addresses of employers, and job titles. If previous job responsibilities are related to the position being sought, they should be described. Summer jobs, as well as part-time jobs held for six months or more, should be listed; any work looks better than no work.

Under "References," list the names, titles, and addresses of three or four people who will testify to your qualifications. You must have permission to use their names, and ideally the list should include people who can vouch for your character, ability as a student, and performance as an employee.

THE INTERVIEW

At an interview, you should be prepared to answer questions about your qualifications and career objectives, but there is no reason for the interview to be one-sided. Interviewers are generally quite willing to discuss how well your goals fit into the firm's future. In addition to specific questions about the position you are seeking, you can inquire about opportunities for advancement, the advisability of continuing your education, or any other matter that concerns you. If the interviewer does not mention salary, you may ask about it; however, most applicants find that firms offer approximately the same salaries and ultimately evaluate them according to other criteria. In short, if you have sufficiently researched potential firms, you know enough to ask the right questions at interviews. While showing your intelligence and interest, you gain the information necessary to make a wise decision.

Jobs are seldom offered at initial interviews, but you should not allow an interview to end until you receive some indication of your standing. A firm can reasonably be expected to set a date for either a second interview or a decision about your employment. If offered a job, you have a similar obligation; you are not expected to make an instant decision, but a date should be set for your response.

FOLLOW-UP LETTER

After an interview with a particularly appealing firm, you can take one more step to distinguish yourself from the competition. Very few applicants write follow-up letters, although it takes only a few minutes to thank the interviewer and express your continued interest in the job:

> Thank you for our interview yesterday. Our discussion of Cranston's growing fluid dynamics' division was very informative, and I am eager to make a contribution to it.
>
> I am looking forward to hearing from you.

The application letter and personal data sheet which follow may be used as models.

7094 Schneider Avenue
Hammond, Indiana 46323
March 3, 1972

Mr. Jeffrey S. Bodge
Personnel Director
Jackson Engineers
1653 Lake Street
Louisville, Kentucky 40214

Dear Mr. Bodge:

Mr. Lyle C. Dawson, Placement Officer at Parker University,
has informed me that your firm is seeking an engineering
graduate with knowledge of surveying and heavy construction.
My four years at Parker have given me thorough training in
these areas.

I will receive my Bachelor of Science Degree in Construction
Technology this June. My courses in the construction field
have covered design, principles, and methods. During the
past four years, my summer and part-time jobs have given me
practical experience in surveying procedures. The enclosed
data sheet provides details about my education and experience.

May I discuss this position and my qualifications at an
interview? My phone number and address are at the top of
the data sheet, and I am available at your convenience.

Yours sincerely,

George R. Kleinfelt

Personal Data Sheet
of
George R. Kleinfelt
7094 Schneider Avenue

INland 7-9265   Hammond, Indiana 46323   March 3, 1972

PERSONAL DETAILS

Age:	22	Draft Status:	I-A: No. 201
Height:	6 ft., 1 in.	Hobbies:	Golf, Hunting
Weight:	175		Auto Mechanics
Health:	Excellent	Membership:	Student Society
Marital Status:	Single		of Engineers

EDUCATION

1968-1972   Parker University
I will receive my B.S. in Construction
Technology in June, 1972. My present
grade point average is 3.46 of 4.00.

Major Courses: Surveying Practice
Boundary Control and
Legal Principles
Route Surveying
Construction Methods
Building Materials

Other Helpful Courses for This Position:
Testing Materials
Design of Structural Steel
Statics
Strength of Materials

1964-1968   Gavit High School, Hammond, Indiana
Graduate, June, 1968

EXPERIENCE

1970-present   Summers and Part-time During School Years
Surveyor's Assistant
Holt Engineering Co., 1711 Marquette St.
Hammond, Indiana 46323

1968-1970   Summers
Estimator
Melton Construction Co., 2534 Cedar Ave.
Highland, Indiana 46322

REFERENCES

Professor Michael R. Lane
Chairman
Construction Technology Dept.
Parker University
Hammond, Indiana 46323

Mr. William A. Holt
President
Holt Engineering Co.
1711 Marquette St.
Hammond, Indiana 46323

Professor Robert T. Marley
Construction Technology Dept.
Parker University
Hammond, Indiana 46323

APPENDIX: COMMON TECHNICAL
WRITING ERRORS

This appendix is essentially a reference or troubleshooting section. It attacks technical writers' most common stylistic and grammatical problems, and emphasizes the objectivity, conciseness, and clarity required in reports. You may, however, want to purchase an English handbook for more comprehensive coverage of grammar.

PERSON

Although the personal pronouns "I" and "we" invariably increase a report's readability, these words are generally not used in technical writing. Report readers expect objective thinking and writing, and they want you to reinforce your role as a detached, scientific observer by avoiding references to yourself. With rare exceptions, you will be expected to say such things as, "The investigation showed that the transistors were faulty," rather than, "My investigation showed. . . ." Also, avoid the word *you* in reports; nobody knows who "you" is.

VOICE

Verbs are either intransitive or transitive. As the word implies, "intransitive" verbs do not transfer action:

Intransitive: The senior engineer arrived.
(The verb *arrived* has no direct object; it completes the action.)

Intransitive: A servomechanism is a feedback system.
(The verb *is* has no direct object; it simply links *feedback system* to *servomechanism*.

Transitive verbs, on the other hand, have a receiver of the action. Transitive verbs are either passive or active:

Passive: The electrode is surrounded by an insulator.
(The subject *electrode* receives the action of the verb.)

Active: An insulator <u>surrounds</u> the electrode.

(The subject *insulator* performs an action. The change from passive to active voice is accomplished by making *insulator* the subject of the sentence rather than the object of the preposition *by;* the previous subject *electrode* becomes the direct object in the active sentence.)

Use the active voice wherever possible in reports. It allows you to write more natural, direct, and concise sentences, and it places the emphasis on the performer of the action rather than the receiver. However, when *you* are the performer of the action, report writing's conventional avoidance of personal pronouns usually forces you to say, "It is recommended . . . ," rather than, "I recommend. . . ."

TENSE

Except for references to past and future events, which demand shifts in tense, use the present tense in technical writing.

Unnecessary Shift: Studies <u>show</u> that the procedure no longer <u>works</u>. It <u>was</u> slower and more time-consuming than the proposed procedure.

Uniform Tense: Studies <u>show</u> that the procedure no longer <u>works</u>. It <u>is</u> slower and more time-consuming than the proposed procedure.

DICTION

Part of your job as a technical writer is to make complex ideas understandable. To help accomplish this, use concrete words and avoid redundancies and wordiness.

CONCRETE WORDS

Given a choice between abstract, highly technical words and simple, concrete, specific words conveying the same meaning, always choose the simple ones; if you must use a potentially confusing word, define it. Jargon, the specialized language used and understood within a department or engineering discipline, should also be avoided in reports which travel beyond the department.

General terms, like *excessive, numerous,* and *frequently,* and words of judgment, like *effective, mediocre,* and *significant,* should be avoided or immediately clarified. They tell the reader very little in comparison to less ambiguous, more specific words.

REDUNDANCY

Redundancy is the repetition of an idea in different words. Some common redundancies and their corrections are listed below:

employed the use of	used
basic fundamentals	fundamentals
completely eliminate	eliminate
alternative choices	alternatives
actual experience	experience
the reason . . . is because	the reason . . . is that
connected together	connected
final result	result
prove conclusively	prove

WORDINESS

From the report reader's standpoint, a good report says what needs to be said and gets it over with. This does not mean that all your reports should be short; however, no report should be longer than it has to be. Words which do not move the report forward are deadwood, needless obstacles to communication. At best they waste your reader's time, and at worst they blur your meaning. Many reports are cluttered by wordy phrases like "at this point in time" [now], and "come in contact with" [contact].

Wordy: Physical equilibrium is a facet of nature that is present in the total environment.

Revised: Physical equilibrium exists everywhere.

Wordy: Experience is relevant to the question as to whether or not an hypothesis is true.

Revised: Experience indicates whether an hypothesis is true.

Wordy: When a body is in static equilibrium, it shows a lack of movement.

Revised: A body in static equilibrium does not move.

Revise the following sentences, removing unnecessary words.

1. Studies of the two units are needed ~~to be conducted~~ to determine which ~~of them~~ would be more economical to purchase.

2. A ~~complete~~ orientation time of about 30 hours is suggested as a sufficient time period for learning the basic and essential operations of the system.

3. The location of the data collector will be in the presently vacated storage room.

4. From answers on the questionnaires it was found that the average amount of use that these units will receive is 100 hours per month.

5. A commonly known fact is ~~that of~~ expansion and contraction of materials, steel being no exception.

6. The alternatives must be examined carefully so that the money spent will be ~~well~~ worth the investment.

7. Inefficiency was ~~said to be the most~~ important factor in the decision to hire a new supervisor.

8. The manufacturing company of this system strongly urges cleaning of the equipment so as to keep intricate parts from every once in a while becoming jammed.

9. Several housing projects are found 2 miles south of Springfield and can be seen to be setting a trend of moving parallel with the highway.

10. Several manufacturers have developed new bumpers to cushion the impact of a car in a collision ~~with another car or~~ in striking an object.

SENTENCES

A sentence is a group of words containing a subject and a verb and expressing a complete thought. Sentences may be composed of smaller subject-verb units called clauses.

MAIN AND SUBORDINATE CLAUSES

Subject-verb units which express complete thoughts are called main, or independent, clauses. If the words forming an independent clause were removed from a sentence, they would form a complete thought by themselves. The two independent clauses in the following sentence are underlined: A 3,000-megawatt nuclear plant requires 1,200 acres, but smaller plants require less land.

Subordinate, or dependent, clauses contain subject-verb units which cannot stand alone; to make sense, they must be linked to independent clauses. The subordinate clause is underlined in the following sentence: Water from nuclear plants must be cooled because it endangers aquatic life. Subordinate clauses serve as nouns, adjectives, or adverbs.

Noun Clause:	The experiment showed that the new material was flawless. (The clause serves as a direct object of the verb *showed*.)
Adjective Clause:	The information which the researchers gathered did not prove anything. (The clause serves as an adjective modifying the noun *information*.)
Adverb Clause:	Sound waves are radiated when the voice-coil assembly is vibrating. (The clause serves as an adverb modifying the words *are radiated*.)

PHRASES

A phrase is a group of related words not containing a subject and a verb. Phrases may serve as nouns, adjectives, or adverbs, and are introduced by prepositions or verbals.

Prepositional Phrases.   In prepositional phrases, the preposition connects a noun or pronoun to the rest of the sentence. Prepositions include the words *with, without, at, on, from, about, by, for, during, in, of, through, to, until,* and *under.* Prepositional phrases serve as either adverbs or adjectives.

The current was connected to the voltmeter.
(The phrase serves as an adverb modifying *was connected.*)

The molecules of the metal were magnetized.
(The phrase serves as an adjective modifying *molecules.*)

Verbal Phrases.   Words derived from verbs and used as nouns, adjectives, or adverbs are called verbals. Gerunds, participles, and infinitives generally join other words to form verbal phrases.

Gerunds always end in *-ing* and serve as nouns:

The system began injecting fuel.
(The phrase serves as the direct object of the verb *began.*)

Participles always serve as adjectives. Present participles end in *-ing* and past participles usually end in *-ed, -en,* or *-t:*

Nothing stopped the men installing the computer.
(The phrase serves as an adjective modifying *men.*)

Built in Texas, prepared by the Wood brothers, and driven by A. J. Foyt, the Coyote-Ford won the race.
(The phrases serve as adjectives modifying *Coyote-Ford.*)

Infinitives are the *to* forms of verbs, although the *to* is sometimes omitted. Infinitives serve as nouns, adjectives, and adverbs:

To begin the project involved many risks.
(The noun phrase serves as the sentence's subject.)

It was a difficult project to begin.
(The adjective phrase modifies the noun *project.*)

The department was eager to begin the project.
(The adverb phrase modifies the verb *was eager.*)

RESTRICTIVE AND NONRESTRICTIVE MODIFIERS

Modifiers are either restrictive or nonrestrictive. Nonrestrictive modifiers add information but do not change, or restrict, a sentence's basic meaning; therefore, they are set off from the rest of the sentence with commas. Restrictive modifiers *do* change the meaning of a sentence; because they are important to the sentence's meaning, they are not set off with commas. Compare the following clauses:

Nonrestrictive: James Jenkins, <u>who was hired in 1970,</u> has been replaced by a machine.

Restrictive: All the equipment <u>that was destroyed</u> must be replaced.

The same modifier may be either restrictive or nonrestrictive depending on your meaning:

Nonrestrictive: The tests, <u>which proved nothing,</u> have been canceled.
(The implication is that *all* the tests have been canceled because they proved nothing.)

Restrictive: The tests <u>which proved nothing</u> have been canceled.
(Here, the implication is that only those tests which proved nothing have been canceled. The clause is restrictive because it restricts the word *tests* to a particular group.)

MISPLACED AND DANGLING MODIFIERS

Sentences become confusing when modifiers do not point directly to the words they modify. Misplaced and dangling modifiers often produce absurd sentences; worse yet, they occasionally result in sentences which make sense, allowing the reader to misinterpret your meaning. Modifiers must be placed in a position that clarifies their relationship to the rest of the sentence.

Misplaced: Before leaving the company, an inspector examines the products carefully.

Correction: An inspector examines the products carefully before they leave the company.

Misplaced: Thermal pollution is worse at nuclear facilities, which must be reduced.

Correction: Thermal pollution, which must be reduced, is worse at nuclear facilities.

PARALLEL STRUCTURE

Sentences are like equations. The structure of a sentence helps clarify your meaning by reinforcing the relationship of your ideas. Elements of a sentence which are connected by coordinating conjunctions (and, but, or, nor, for, yet, so) carry the same weight and must be grammatically similar, or parallel. Items in a series of words, phrases, or subordinate clauses have the same value, and their grammatical structure must indicate their equality. If coordinate elements in a sentence do not have a parallel pattern, the sentence becomes awkward and potentially confusing.

Faulty: Management guarantees that the old system will be replaced and to consider the new proposal.
(The subordinate clause and infinitive phrase are not in parallel structure.)

Parallel: Management guarantees that the old system will be replaced and that the new proposal will be considered.

Faulty: Use the cylinder with a diameter of 3³⁄₁₆ inches and 1½ inches high.
(To be parallel, the two items in this series should be objects of the preposition *with*.)

Parallel: Use the cylinder with a diameter of 3³⁄₁₆ inches and a height of 1½ inches.
(Parallelism can also be achieved by making both items objects of the verb *has*.)

Parallel: Use the cylinder which has a <u>diameter of 3³⁄₁₆ inches</u> and a <u>height of 1½ inches.</u>

Faulty: A successful firm is capable of <u>manufacturing a product,</u> <u>marketing it,</u> and <u>make a profit.</u>
(For this series to be parallel, all three objects of the preposition *of* must be participles.)

Parallel: A successful firm is capable of <u>manufacturing a product,</u> <u>marketing it,</u> and <u>making a profit.</u>

Revise the following sentences, making their coordinate elements parallel.

1. A proposal for the use of the land and how to finance the project has been submitted.

2. The manual contains directions for excavating earth, laying asphalt, and then installation of synthetic turf.

3. Steps must be taken to remove phosphates from waste material, devising a system for early detection of oil spills, and creation of laws to prohibit the dumping of industrial waste.

4. The most frequent errors are those caused by improper calibration or that are the result of careless operation.

5. A monthly charge of $400 is for computer time, printed output, and to maintain the system.

6. The measuring device's advantages are its reduction of errors, requiring less maintenance, and provides greater accuracy.

7. The shaft, with a diameter of ⅜ inch and 1¼ inches long, is grooved toward the end.

8. A surveying instrument should be purchased not only because it reduces human errors but also resulting in saved time.

9. A shaft extends from the socket holder, goes through the ratcheting gear, and to a push button.

10. The Wankel, a thoroughly tested engine and used by Japanese and German auto manufacturers, will cost less when full production begins.

SUBJECT-VERB AGREEMENT

The subject and verb of a sentence must both be singular or plural. Almost all problems with agreement are caused by failure to correctly identify the subject.

Faulty: The stockpile of chemicals are located in an uninhabited area.
(The subject *stockpile* is singular; *chemicals* is the object of a preposition.

Correction: The stockpile of chemicals is located in an uninhabited area.

Faulty: The distances that E and W move around the fulcrum determines the amount of effort needed.

Correction: The distances that E and W move around the fulcrum determine the amount of effort needed.

Faulty: The committee are investigating the sales report.
(When a *collective noun* refers to a group as a unit, the verb must be singular. Other collective nouns are *management, union, team, audience* and *jury*.

Correction: The committee is investigating the sales report.

Faulty: Each of the steelworkers are highly skilled.
(Indefinite pronouns, such as *each, everyone, either, neither, anyone* and *everybody*, take a singular verb.

Correction: Each of the steelworkers is highly skilled.

Faulty: Neither the foreman nor the laborers wants a strike.
(When compound subjects are connected by *or* or *nor*, the verb must agree with the nearer noun.

Correction: Neither the foreman nor the laborers want a strike.

The word *data* causes some confusion for technical writers. Traditionally, the word has been considered a plural, but the trend is toward viewing it as a singular, in the sense of "information." Today, you may use either a singular or plural verb with *data*, but be consistent throughout the report.

Singular: The data is. . . . It. . . . This data. . . .

Plural: The data are. . . . They. . . . These data. . . .

Select the correct verbs in the following sentences.

1. Cable for the zoom, pan, and tilt mechanisms (cost, costs) $60 per 1,000 feet.

2. The piston engine now used by all U. S. manufacturers has been developed to the point where further reductions in pollution (seem, seems) unlikely in the near future.

3. Following the first section of the report (is, are) descriptions of the system and its installation.

4. None of the employees (is, are) going to like the revised vacation schedule.

5. The function of the cables (is, are) to provide voltage required by the zoom lens.

6. These facts about water pollution in Lake Michigan (was, were) gathered from an article about the problem.

7. For some operations with a slide rule, the choice of scale combinations (make, makes) no difference.

8. The head is made of chrome-plated, drop-forged steel which (encase, encases) the ratchet and socket-releasing mechanisms.

9. The cables which connect the camera to a monitor (carry, carries) the video signal.

10. Neither the surveyor nor the engineer (perform, performs) routine operations in the field.

PRONOUN REFERENCE

A pronoun must refer directly to the noun it stands for, its antecedent. Pronouns commonly used in technical writing include *it, they, who, which, this* and *that*. Never use *they* as an indefinite pronoun; when you write, "They say that . . . ," make sure your reader knows who "they" are. *It* may be used sparingly as an indefinite pronoun ("It is obvious that. . . ."), but overuse of the indefinite *it* leads to confusion.

As in subject-verb agreement, a pronoun and its referent must both be singular or plural. Collective nouns generally take the singular pronoun *it* rather than *they*.

Problems result when pronouns such as *they, this,* and *it* are used carelessly, forcing the reader to figure out their referents:

Vague: Research teams examined sites for a new chassis factory. They overlooked the lake front.

Clear: Research teams examined sites overlooking the lake front for a new chassis factory.

Vague: The consultants recommended a new method for selling aluminum. This is the company's best bet for the future.

Clear: The consultants recommended a new method for selling aluminum, the company's best bet for the future.

Vague: The Atomic Energy Commission determines the criteria for selecting a nuclear site. It includes a specified distance from high-population zones.

Clear: The Atomic Energy Commission, which determines the criteria for selecting a nuclear site, specifies that the site be a certain distance from high-population zones.

Revise the following sentences, making the pronoun references unmistakable.

1. The faster test drivers drove the Wankels, the quieter they got.

2. Federal officials say that auto makers must adhere to new laws restricting auto emissions. They brought them upon themselves.

3. The company had high hopes for a new research program, but it encountered financial problems.

4. Asphalt will be spread over the field as soon as excavation is completed. This takes approximately 5 weeks.

5. He will explain how the new component can be adapted to the present system as well as its basic principles.

6. During the compression stroke, the piston moves up, compressing the mixture. Then the mixture ignites, and expanding gas forces it down.

7. The field can be covered during the football season, which requires approximately 45 minutes.

8. Several companies are competing with General Motors and Ford, so they must start their own Wankel programs.

9. In the operation of an automatic cigarette lighter, the spring around the shaft is compressed until it touches the bottom of the base.

10. When the foreman read a report about personnel sleeping on the job, he had no alternative but to consider it carefully.

COMMAS, COLONS, SEMICOLONS

The function of punctuation is to help clarify the meaning of your sentences. A punctuation mark as seemingly unimportant as a comma can radically change your reader's interpretation of a sentence. This section explains the basic uses of the comma, colon, and semicolon.

The Comma. Use commas in the following ways:
1. To separate two main clauses connected by a coordinating conjunction (and, but, or, nor, for, yet, so). The comma may be omitted if the clauses are very short.

Radar feeds the information back to the control station, and a new course is relayed to the missile.

2. To separate introductory subordinate clauses or phrases from the main clause.

Clause: When the target changes course, radar detects the change.

Phrase: Sensing a reduction in pressure, the regulator sends more gas into the pipeline.

3. To separate words, phrases, or clauses in a series.

Words: The recorder contains a noise-reduction unit, a frequency-equalizer unit, and a level-control unit.

Phrases: The skill is needed in industry, in education, and in government.

Clauses: Select equipment that has durability, that requires little maintenance, and that the company can afford.

4. To set off nonrestrictive appositives, phrases, and clauses. (Dashes and parentheses also serve this function. Parentheses may be used frequently, but use dashes sparingly.)

Appositive: Clem Stacy, the foreman, fixed the generator.

Phrase: The engine, beginning its seventh year of service, finally needed maintenance.

Clause: The system, which is somewhat complicated, requires calibration at various intervals.

5. To separate coordinate but not cumulative adjectives.

Coordinate: He rejected the distorted, useless recordings. (The adjectives are coordinate because they modify the noun independently. They could be reversed with no change in meaning: useless, distorted recordings.)

Cumulative: An acceptable frequency-response curve was achieved. (The adjectives are not separated by commas because they modify intervening adjectives as well as the noun. Cumulative adjectives cannot be reversed without distorting the meaning: frequency-response acceptable curve.)

6. To set off conjunctive adverbs and transitional phrases.

Conjunctive Adverbs: The new regulations, however, caused a morale problem.

The crane was very expensive; however, it paid for itself in 18 months.

Therefore, the branch plant should not be built until 1976.

Transitional Phrases: On the other hand, the maintenance crew operated efficiently.

Workmanship on Mondays and Fridays, for example, is far below average.

The Colon.   Use colons in the following ways:
1. To separate an independent clause from a list of supporting statements or examples.

A piston has the following strokes: intake, compression, power, and exhaust.

2. To separate two independent clauses when the second clause explains or amplifies the first.

The Wankel engine is potentially better than conventional engines for one major reason: it has 40 percent fewer parts.
(Do not capitalize the first word of the second independent clause unless you want to give the clause special emphasis.)

The Semicolon.   Use semicolons in the following ways:
1. To separate independent clauses not connected by coordinating conjunctions (and, but, or, nor, for, yet, so).

The storm stopped the surveyors; in fact, it stopped all work at the site.

2. To separate independent clauses connected by coordinating conjunctions *only* if the clauses are long or have internal punctuation.

In a gunfire-control system, the target plane is moving, and the input information is variable, depending on the plane's speed and range; but radar, acting as the sensing device, feeds the input information to the control station.

3. To separate independent clauses when the second one begins with a conjunctive adverb (therefore, however, also, besides, consequently, nevertheless, furthermore).

The machine performs better than all the others; therefore, it should be purchased.

4. To separate items in a series if the items have internal punctuation.

Plants have been proposed for Kansas City, Missouri; Seattle, Washington; and Orlando, Florida.

NONSENTENCES

Comma splices, run-on sentences, and sentence fragments are among the most serious sentence problems. This section shows how to recognize them and correct them.

COMMA SPLICES

A comma splice occurs when two independent clauses are connected, or spliced, with only a comma. You can correct comma splices in four ways:

1. Replace the comma with a period to separate the two sentences.

Splice: Friction results from the movement of one material across another, uneven surfaces cause greater friction.

Correction: Friction results from the movement of one material across another. Uneven surfaces cause greater friction.

2. Replace the comma with a semicolon *only* if the sentences are very closely related.

Splice: An hypothesis is an assumption, therefore it must be tested.

Correction: An hypothesis is an assumption; therefore, it must be tested.
(The word *therefore* is a conjunctive adverb. When you use a conjunctive adverb to connect two sentences, always precede it with a semicolon and follow it with a comma. Other conjunctive adverbs are *however, also, besides, consequently, nevertheless,* and *furthermore.*

3. Insert a coordinating conjunction after the comma, making a compound sentence.

Splice: The gas-air mixture burns in internal-combustion engines, it does not explode.

Correction: The gas-air mixture burns in internal-combustion engines, but it does not explode.
(The words *and, but, or, nor, for, yet,* and *so* are coordinating conjunctions.)

4. Subordinate one of the independent clauses by beginning it with a subordinating conjunction or a relative pronoun. Do not use this method for correcting a comma splice unless the clause *should* be given less emphasis.

Splice: The operation of a radar system includes three main sequences, they are the generation, transmission, and reception of a signal.

Correction: The operation of a radar system includes three main sequences, which are the generation, transmission, and reception of a signal.
(The words *which, that, who,* and *what* are relative pronouns. Frequently used subordinating conjunctions are *where, when, while, because, since, as, until, unless, although, if,* and *after.*)

Using the methods shown above, correct the following comma splices.

1. The condition, which steadily grows worse, will continue to deteriorate, action to alleviate the pollution problem, including the creation of new laws, is badly needed.

2. During the first year, the company saved $20,000, these savings resulted from better inventory control.

3. Several antennas have been considered, one of them, the "big wheel" antenna, satisfies both criteria.

4. The system operates simply and efficiently, also it keeps track of the current progress of the items.

5. When multiplying with the slide rule, the result is shown at the right index if the slide is moved to the left, however, the result is shown at the left index if the slide is moved to the right.

6. As the temperature fluctuates, so does the length of the tape, if this is not taken into consideration, precision suffers.

7. A carburetor's operation is based on the Venturi principle, that is, a gas or liquid flowing through a restriction will increase in speed and decrease in pressure.

8. Two valves control fluid flow, one releases pressure in the system and allows the ram to retract into the cylinder, the second controls the pistons.

9. When the iron becomes magnetized, it can attract other pieces of iron, when no current is flowing in the coiled wire, the iron is not magnetized.

10. Consequently, the Chicago area will need a new airport, in fact, construction should be started immediately.

RUN-ON SENTENCES

Run-on, or fused, sentences look like comma splices without the comma. The independent clauses are run together with no punctuation between them. To eliminate fused sentences, use one of the four methods explained above: (1) place a period between the two clauses, (2) place a semicolon between them, (3) place a comma and a coordinating conjunction between them, (4) or place a relative pronoun or subordinating conjunction between them.

Using the methods above, correct the following run-on sentences.

1. Jerry Chapetta's article is entitled "Great Lakes: Great Mess" it appears in the May, 1968 issue of *Audubon*.

2. According to Chappetta, the chemical symbol for water in the Great Lakes should be H_2Os_{10} this symbol stands for 2 parts hydrogen, 1 part oxygen, and 10 parts stupidity, greed, neglect, and mismanagement.

3. During the installation period, business operations will not be disrupted switching over to the new system will take less than a day.

4. When the level of the capsule changes, the heater turns off the change causes mercury to roll away from the contacts.

5. The function of the carburetor is to atomize the fuel and mix it with air flowing into the engine the carburetor must also meter the fuel to provide the proper fuel-air ratio.

6. An engine converts burning fuel into usable energy the energy is in the form of a rotating shaft.

7. The temperature in any part of the building can be controlled this is beneficial during periods of reduced occupancy.

8. Careful driving is important but the automobile should be designed for safety many people are more concerned about the auto's appearance.

9. The early air bags lacked a reliable sensor a sensing device had to be developed.

10. Lack of funds is the most important problem in pollution control for several years the government has promised assistance but very little has been delivered.

SENTENCE FRAGMENTS

Sentence fragments are incomplete thoughts which have been punctuated as complete sentences. Fragments are often subordinate clauses, prepositional phrases, and verbal phrases. As the following examples show, they must be connected to the preceding or following sentence to gain meaning.

1. Connect subordinate clauses to independent clauses.

Fragment: The company continues to lose money. <u>Although production has increased.</u>
(The fragment is a subordinate clause beginning with
a subordinating conjunction. Other subordinating conjunctions are *where, when, while, because, since, as,
until, unless, if,* and *after.*)

Correction: The company continues to lose money although production has increased.

Fragment: The problem originates in the transformer. <u>Which
does not provide enough electric energy.</u>
(The fragment is a subordinate clause beginning with
a relative pronoun. Other relative pronouns are *who,
that,* and *what.*)

Correction: The problem originates in the transformer, which does
not provide enough electric energy.

2. Connect prepositional phrases to independent clauses.

Fragment: Civil engineering requires many skills. <u>For example,
drafting and surveying.</u>
(The fragment is a prepositional phrase. Other prepositions are *with, without, at, on, from, about, by, during, in, of, through, to, until,* and *under.* The fragment
can be converted into a subordinate clause, as in the
first example below, or made into a participial phrase.

Correction: Civil engineering requires many skills, such as drafting and surveying.

Correction: Civil engineering requires many skills, including drafting and surveying.

3. Connect verbal phrases to independent clauses.

Fragment: Kinetics is a field of dynamics. <u>Consisting of all aspects of motion.</u>
(Verbal phrases, including this participial phrase

modifying *field*, often begin with *-ing* words. Such phrases must be linked to independent clauses.)

Correction: Kinetics is a field of dynamics consisting of all aspects of motion.

Fragment: The writer studied the statistics. <u>To insure his accuracy.</u>
(Infinitive phrases, including this one modifying *studied*, often begin with *to* and a verb. They must be linked to independent clauses.)

Correction: The writer studied the statistics to insure his accuracy.

Using the methods shown above, correct the following sentence fragments.

1. Although these experiments are performed by experienced people at the laboratory.

2. Man, the cause of all pollution in Lake Michigan and the only hope for saving the lake.

3. When atoms absorb the spark's energy and give off light.

4. Material presented to emphasize the fate that seems to be awaiting Lake Michigan.

5. The offending companies, having been named in the report along with evidence of their guilt.

6. Five elements are checked immediately. The other four sent to the laboratory for testing.

7. That the high-voltage system, being more efficient, causes fewer technical problems than other systems.

8. An electronic system has the same basic problems as a mechanical system; speed and accuracy.

9. Because the manual method of testing requires at least 7 minutes regardless of the employee's skill.

10. Each year more than $10,000 can be saved if the company buys a new measuring system. No decrease in quality if the measurement is accurate.

PARAGRAPHS

A paragraph consists of several sentences anchored by a topic sentence. The topic sentence expresses the paragraph's central idea, and the remaining sentences develop, explain and support the central idea.

TOPIC SENTENCES

The central idea of each paragraph must be stated in a strong topic sentence. In technical writing, the topic sentence usually appears at the beginning of a paragraph, followed by details which support and clarify the central idea. This deductive structure gives your paragraphs the direct, straightforward style preferred by most report readers. Inductive paragraphs, which may occasionally be used, begin with the details and build up to a topic sentence at the end of the paragraph.

All deductive paragraphs follow a statement-support pattern. The topic sentence consists of a generalization which must be explained or illustrated with details, as in the following example:

Uranium will continue to be the major fuel for nuclear reactors. The latest statistics show that enough uranium is already available to supply the needs of the United States beyond 1990. The supply of uranium will increase as technology develops new methods of locating, mining, and processing it. As the supply increases, uranium's cost should decrease, making it even more attractive as a fuel for nuclear reactors.

Two types of statement-support paragraphs, effect-cause and comparison-contrast, are common in technical reports, Whichever pattern you use, insure that your topic sentence clearly fixes the direction and boundary of the paragraph.

EFFECT-CAUSE

Begin an effect-cause paragraph by stating an effect; then explain, or trace, its causes:

A coal shortage is developing in the United States, not because of the lack of coal, but because of decreased coal mining. Previously, the coal industry was plagued with overproduction, but mechanization has greatly increased the cost of opening new mines and dampened the enthusiasm of investors. Labor problems and the necessity for better health and safety equipment have also discouraged investors. Recent restrictions on the sulfur content of coal have increased the shortage; the industry now has difficulty mining enough coal of the quality required by the government to supply growing needs.

COMPARISON-CONTRAST

In paragraphs involving alternatives, the comparison-contrast pattern allows you to present data for the alternatives and to discuss each alternative's advantages and disadvantages:

Operation and maintenance costs are lower for nuclear plants than for coal-fired plants. Nuclear plants require no equipment for handling and burning coal, which results in fewer employees and smaller payrolls. A coal plant needs approximately 250 people, but a nuclear facility can operate efficiently with approximately 100 workers. Therefore, the cost of running a coal plant is about 0.24 mills per kilowatt-hour, versus 0.10 mills for a nuclear station.

TRANSITION

Transition between and within paragraphs causes no problems if you present ideas in a logical and orderly way. Synonyms and repetition of key words achieve transition as you progress from idea to idea ("The company won 15 contracts last year. . . . These victories. . . . Most of the contracts. . . ."). To further clarify the relationship of your ideas, use connective words and expressions:

for example	on the other hand
for instance	on the contrary
in fact	therefore
in other words	furthermore
in short	consequently
in addition	similarly
however	in summary
nevertheless	in conclusion

CAPITALIZATION

The conventional rules of capitalization apply to technical writing. Within each report, capitalize uniformly and remember that the trend in industry is away from overcapitalization. For example, say, "The senior project engineer, Mr. Reynolds, . . ." rather than capitalizing each word of his title, and say, "Tinius Olsen testing machine," rather than capitalizing the name of the product as well as the trademark.

Use capitalization to provide emphasis and to indicate levels of headings within reports. Titles of reports, titles of visuals, and first- and second-level headings can be entirely capitalized. Capitalize the first word and each succeeding word (except prepositions, conjunctions, and articles) of third- and fourth-level headings and labels on visuals. Also, capitalize references to visuals ("As Figure 4 shows, . . .").

ABBREVIATIONS

Use abbreviations only for long words or combinations of words which must be used several times in a report. For example, if words like *Fahrenheit, cubic inches, pounds per square inch,* or *British thermal units* must be used several times in a report, abbreviate them to save space. However, short words like *acre* or *ton* should not be abbreviated. Most authorities say that you should write *in.* rather than *inch,* but you have to abbreviate *inch* quite often to significantly shorten a report. The following are four rules for abbreviating.

1. If the abbreviation of a long word or group of words might confuse your reader, spell out the word in parentheses the first time you use its abbreviation.
2. Use small letters except for the abbreviation of proper nouns such as *British thermal units* (Btu).
3. Do not add "s" to form the plural of an abbreviation.
4. Do not use abbreviations for units of measurement preceded by approximations. You may say "15 psi" but not "several psi."

POSSESSIVES

The following are four basic rules for showing possession:

1. Add an apostrophe and an "s" to show the possession of nouns which do not already end with "s."

   a corporation's profits
   the foreman's orders
   the foremen's orders

2. Add only an apostrophe to plural nouns ending with "s" (eleven corporations' profits).
3. For singular nouns of one syllable ending in "s," add an apostrophe and an "s" (boss's orders; punch press's cost) but add only an apostrophe if the singular noun has two or more syllables (Mark Williams' job).
4. Do not add an apostrophe to personal pronouns (theirs, ours, its). The only time *its* needs an apostrophe (it's) is when it's a contraction for the words *it is*.

HYPHENS

Use hyphens to connect two or more modifiers that express a unified, or one-thought, idea (high-frequency system; alternating-current motor). If some of the following modifiers were not hyphenated, confusion would result:

   energy-producing cells
   eight-hour shifts
   15 eight-cylinder, 300-horsepower engines
   cement-like texture
   A-frame construction
   plunger-type device
   trouble-free system

The following are four more rules for hyphenating:

1. Use hyphens to connect compound nouns used as units of measurement (kilowatt-hour).

2. Do not hyphenate adverb-adjective combinations (recently altered system).

3. Use hyphens to connect whole numbers and typewritten fractions (six 1-1/2-horsepower motors) and to connect compound numbers from 21 through 99 when they are spelled out (Thirty-five safety violations were reported.).

4. Use suspended hyphens for a series of adjectives that you would ordinarily hyphenate (10- , 20- , and 30-foot beams).

NUMBERS

Because numbers are used so often in technical writing, rules for their usage are somewhat different than for other kinds of writing. The following rules cover most situations, but when in doubt whether to use a figure or a word, remember that the trend in report writing is toward using figures.

1. For numbers accompanying units of measurement, use figures:

   1 gram
   33 1/3 percent
   8 yards
   0.452 minute
   $6.95
   5 weeks

2. When not used with units of measurement, spell out numbers below ten, and use figures for ten and above:

   four cycles
   three-fourths of the union
   10 machines
   1,835 members

3. When numbers above and below ten are in a series, use figures for all of them.

4. Use figures for approximations (around 175 pounds); until recently, approximations were spelled out.

5. Spell out numbers that begin sentences.

6. For compound-number adjectives, spell out the first one or the shorter one to avoid confusion (75 twelve-volt batteries).
7. For numbers above a million, use a combination of figures and words (2 million miles).
8. Place the last two letters of the ordinal after fractions used as nouns (1/50th of a second), but not after fractions that modify nouns (1/50 horsepower). Spell out ordinals below ten (third mile, ninth man), but for ten and above, use the number and the last two letters of the ordinal (10th day, 21st year).

SYMBOLS

Like abbreviations, symbols save space but cause confusion. Except for the symbol for *dollars*, symbols should not be used within the paragraphs of a report. To save space in tables and drawings, however, use simple symbols like the ones for *feet, inches* and *degrees*. Use highly technical symbols in the report's appendix if you are certain that they will be understood.

INDEX